# Lecture Notes on Data Engineering and Communications Technologies

Volume 174

The aim of the book series is to present cutting edge engineering approaches to data technologies and communications. It will publish latest advances on the engineering task of building and deploying distributed, scalable and reliable data infrastructures and communication systems.

The series will have a prominent applied focus on data technologies and communications with aim to promote the bridging from fundamental research on data science and networking to data engineering and communications that lead to industry products, business knowledge and standardisation.

Indexed by SCOPUS, INSPEC, EI Compendex.

All books published in the series are submitted for consideration in Web of Science.

Isaac Woungang • Sanjay Kumar Dhurandher
Editors

# 5th International Conference on Wireless, Intelligent and Distributed Environment for Communication

## WIDECOM 2022

Springer

*Editors*
Isaac Woungang
Department of Computer Science
Ryerson University
Toronto, ON, Canada

Sanjay Kumar Dhurandher
Department of Information Technology
Netaji Subhas University of Technology
New Delhi, Delhi, India

ISSN 2367-4512 ISSN 2367-4520 (electronic)
Lecture Notes on Data Engineering and Communications Technologies
ISBN 978-3-031-33241-8 ISBN 978-3-031-33242-5 (eBook)
https://doi.org/10.1007/978-3-031-33242-5



This Springer imprint is published by the registered company Springer Nature Switzerland AG
The registered company address is: Gewerbestrasse 11, 6330 Cham, Switzerland

# Welcome Message from WIDECOM 2022 General Chair

Welcome to the 5th International Conference on Wireless, Intelligent, and Distributed Environment for Communication (WIDECOM 2022).

The last decade has witnessed tremendous advances in computing and networking technologies, with the appearance of new paradigms such as Internet of Things (IoT) and cloud computing, which have led to advances in wireless and intelligent systems for communications. Undoubtedly, these technological advances help improve many facets of human lives, for instance, through better healthcare delivery, faster and more reliable communications, significant gains in productivity, and so on. At the same time, the associated increasing demand for a flexible and cheap infrastructure for collecting and monitoring real-world data nearly everywhere, coupled with the aforementioned integration of wireless mobile systems and network computing, raises new challenges with respect to the dependability of integrated applications and the intelligence-driven security threats against the platforms supporting these applications. The WIDECOM conference is a conference series that provides a venue for researchers and practitioners to present, learn, and discuss recent advances in new dependability paradigms, design, and performance of dependable network computing and mobile systems, as well as issues related to the security of these systems.

Every year, WIDECOM receives several submissions from around the world. Building on the success from last year, WIDECOM 2022 presents an exciting technical program that is the work of many volunteers. The program consists of a combination technical papers, keynotes, tutorials, and posters. The technical papers are peer reviewed by program committee members who are experts in their fields, through a blind process.

We received a total of 32 papers this year, and accepted 12 papers for inclusion in the proceedings and presentation at the conference, which corresponds to an acceptance rate of 37.5%. Papers were reviewed by three PC members.

WIDECOM 2022 is privileged to have select guest speakers to provide stimulating presentations on topics of wide interest. This year's Distinguished Speakers are

- Professor Xavier Fernando, Director of Toronto Metropolitan University Communications Lab, Department of Electrical and Computer Engineering, Toronto Metropolitan University, Toronto, Canada, IEEE Communications Society Distinguished Lecturer, and IEEE Canada Vitality Coordinator.
- Professor Abdelwahab Hamou-Lhadj, Director of Software Research and Technology Lab, Department of Electrical and Computer Engineering, Gina Cody School of Engineering and Computer Science, Concordia University, and OMG-Certified UML Professional and OMG-Certified Expert in BPM Certification Programs.

We would also like to thank our tutorial presenters of this year:

- Dr. Wei Lu, Department of Computer Science, Keene State College, The University System of New Hampshire, USA.
- Dr. Glaucio H. S. Carvalho, Department of Computer Science and Engineering, Brock University, Ontario, Canada.
- Dr. Simon Chege, School of Electrical, Electronic and Computer Engineering, University of KwaZulu-Natal, South Africa.

We would like to thank all of the volunteers for their contributions to WIDECOM 2022. Our thanks also go to all the authors, and our sincere gratitude goes to the Program Committee, who carefully review the submissions.

We would also like to thank the organizing committee and our sponsors:

- The Department of Computer Science, University of Windsor.
- Springer LNDECT Series, for publishing the Conference Proceedings.

Finally, we thank all the attendees and WIDECOM 2022 community for their continuing support, by submitting papers and volunteering their time and talent in many ways.

We hope you will find the papers interesting and will enjoy the conference.

**WIDECOM 2022 Conference Chair**

Dr. Sherif Saad Ahmed

# Welcome Message from the WIDECOM 2022 Program Chair

Welcome to the 5th International Conference on Wireless, Intelligent, and Distributed Environment for Communication (WIDECOM 2022), which will be held on October 12–14, 2022, at the University of Windsor, Windsor, Ontario, Canada.

WIDECOM 2022 provides a forum for researchers and practitioners from industry and government to present, learn, and discuss recent advances in new dependability paradigms, design, and performance of dependable network computing and mobile systems, as well as issues related to the security of these systems.

The papers selected for publication in the proceedings of WIDECOM 2022 span many research issues related to the aforementioned research areas, covering aspects such as algorithms, architectures, protocols dealing with network computing, ubiquitous and cloud systems and Internet of Things systems, integration of wireless mobile systems and network computing, and security. We hope the participants to this conference will benefit from this coverage of a wide range of current hop-spot related topics.

In this edition, 32 papers were submitted, and peer-reviewed by the Program Committee members and external reviewers who are experts in the topical areas covered by the papers. The Program Committee accepted 12 papers (about 37.5% acceptance ratio). The conference program also includes two distinguished keynote speeches and three tutorials.

Our thanks go to the volunteers who have contributed to the organization of WIDECOM 2022. We would like to thank all authors for submitting their papers. We would also like to thank the Program Committee members for thoroughly reviewing the submission and making valuable recommendations. We would like to thank the WIDECOM 2022 Local Arrangement team for the excellent organization of the conference, and for their effective coordination creating the recipe for a very successful conference.

We hope you will enjoy the conference.

**WIDECOM 2022 Program Committee Chair**

Dr. Sanjay Kumar Dhurandher, Netaji Subhas University of Technology, New Delhi, India

# Contents

# Chapter 1
# Efficient Fake News Detection Method Using Feature Reduction

**Rayhaan Pirani and Ehsan Ur Rahman Mohammed**

## 1.1 Introduction

Fake news has traditionally been defined as propaganda. People in power primarily used it to sway public opinion toward a specific view by propagating falsehoods. Fake news can be incredibly damaging to society. It may be used to modify people's beliefs and perceptions about specific phenomena, resulting in their change of behavior. Such misinformation may result in people making decisions detrimental to their well-being, such as refusing to accept treatment or medication intended for a particular ailment.

Identifying fake news articles can be difficult for many people because such articles are intended to read like original news articles. Many people fail to verify sources or consider alternative possibilities for two key reasons: the desire for instant gratification and their trust in news sources shared by their family and friends. Algorithms to identify and warn if a text contains fake news can be used to help people avoid potential misinformation and aid them in finding unbiased news sources instead. This can also lead to a reduction in the spread of such fake news and discourage the creation of such articles. It is imperative to ensure that fake news articles are detected as soon as possible and that people accessing them are warned about potential misinformation to make an informed decision to trust the sources they access. For example, when a user is unsure of the credibility of a news article,

Rayhaan Pirani and Ehsan Ur Rahman Mohammed contributed equally.

R. Pirani · E. U. R. Mohammed (✉)
School of Computer Science, University of Windsor, Windsor, ON, Canada
e-mail: piranir@uwindsor.ca; rahman6a@uwindsor.ca



I. Woungang, S. K. Dhurandher (eds.), *5th International Conference on Wireless, Intelligent and Distributed Environment for Communication*, Lecture Notes on Data Engineering and Communications Technologies 174,
https://doi.org/10.1007/978-3-031-33242-5_1

the user can use fake news detection to know whether an article is a fake or real news.

With research carried out in various subdomains of natural language processing (NLP) and its varied application areas, we can state unequivocally that the race is for better performance based on evaluation metrics such as accuracy, computational burden, and storage requirements. The other significant contribution of researchers across artificial intelligence domains is making neural network models more explainable and interpretable.

This study used two fake news detection datasets to experiment on a proposed solution. The dataset has to be preprocessed to make training efficient and fast; in addition, the dataset needs to be understood using the exploratory analysis. We then pass the dataset to reduce its dimensionality. Unsupervised approaches such as clustering are often used for feature reduction; thus, the present study employed this approach. Specifically, we use singular-valued decomposition (SVD) clustering because of the identified benefits, such as the option to select a custom number of reduced features. Then, the obtained features of the individual articles are passed to a Recurrent Neural Network (RNN) model to predict whether the news item is authentic or made up.

### *1.1.1 History of Fake News Detection*

Bondielli and Marcelloni [1] surveyed studies on fake news published since 2006. It highlights that the research on fake news and its detection took off after the 2016 US presidential election. The survey also states that as the research on fake news is new; no benchmark datasets are available yet. An overview of various techniques, especially the machine learning (ML) and deep learning (DL) ones, were briefly highlighted. The paper's concluding remark also stated that DL methods perform better compared to the rest. The survey [2, 3] defines the concept of fake news in the context of social sciences by reviewing the methods that detect fake news using four perspectives, namely, their content, writing style, propagation pattern, and credibility of their source. The challenges in current fake news research were also highlighted. The survey [4] comprehensively reviews various NLP and DL approaches to fake news detection using datasets like Facebook Hoax, LIAR, and FakeNewsNet. LSTMs and ensemble models gave high accuracy, and future research in these methods was encouraged in the paper [4].

The general observable trend clearly states that each approach for fake news detection has its demerits. For instance, deep learning methods are black-box methods, so they are uninterpretable; they are data-hungry models too. Machine learning methods are slower and have reduced the ability to learn complex data patterns. Our work tries to bridge the gap by using the best of both worlds. We use machine learning to reduce data and alleviate the problem of data hunger, and we use a deep learning model to learn complex patterns that are a prominent feature of fake news articles. Central to our study is [5], which helps us weigh the performance of DL and ML approaches for fake news and aids in our understanding that a middle way is better.

Our focus is on formal news articles because the credibility of social media posts by default is low. Next, the processing of social media posts takes a considerable amount of time and requires various other techniques from other related computer science domains, such as social network analysis. Lastly, dealing with social media content always demands maintenance of users' data privacy and integrity. Ensuring data privacy and integrity are significant challenges that many NLP researchers pursue. In this paper, our target is to understand the achievability of our proposed work and its consequences; hence, pursuing social media posts in our future works was considered more relevant.

### *1.1.2 History of Feature Reduction*

Feature reduction [6] (FR) dates back to 1901 with the emergence of principal component analysis (PCA). PCA can be described as an analytic algorithm often used to translate high-dimensional data to lower dimensions. The initial motivation behind FR was the scientific community's desire to fit higher-dimensional data to a straight line or plane. FR methods utilize concepts such as projections, factorization, and quantization, which are central to the mathematical domain of linear algebra. FR continues to be vital due to big data, especially of the kind where variables are more than observations, such as in genome sequencing and text analysis. FR has touched and has become integral to all domains of data analysis, for instance, image compression, time series analysis, and search engine optimization.

## 1.2 Related Works

To the best of our knowledge, there were no studies analyzing the efficiency of fake news detection methods or the efficiency and effectiveness trade-off of these methods. The work presented in [7] is one of the first research works utilizing an FR technique in fake news detection. The authors use K-means clustering on the raw fake news articles and pass the reduced dataset into a Support Vector Machine-based (SVM) classifier for training. The work effectively reduced the training time but could not give novel results in terms of accuracy, most likely due to the classifier being a machine learning rather than a deep learning one.

The motivation to use FR as a preprocessor for a DL-based classifier comes from [8–10]. Survey [8] performs segmentation of license plate in an image using K-means and connected component analysis. The segmented part is passed for character recognition using a convolutional neural network (CNN). Survey [10] is a parallel drawn in NLP using IsoMap FR to perform text classification using SVMs.

Lastly, we take inspiration from [9], where the authors proposed a model with an FR method based on DL and a classifier based on ML, as opposed to our approach. The classifier being an ML-based method tends to be suboptimal. Machine learning prediction techniques were used to predict protein classes in the study [11]. The authors focus on comprehending various protein families and classifying the protein family type sequence. Machine learning algorithms Naive Bayes (NB) and Random Forest (RF) with count vectorized feature and LSTM are used. The LSTM algorithm outperforms the RF and NB algorithms in predicting protein class, which motivates us to use something similar for the task of fake news classification.

### *1.2.1 Traditional Machine Learning Approaches*

**For Formal News Articles** The authors in [12] deploy XGBoost to do fake news detection on formal news articles in Indonesian using a proprietary dataset collected from various Indonesian news websites. The news items were tokenized and converted into TF-IDF vectors to be fed into the classifier. XGBoost is the top-performing classifier among all the other ML models, and the authors leverage this fact and achieve an accuracy of more than 92%. The authors of [13] describe an approach for detecting fake news using a random forest classifier and NLP. A dataset was constructed by collecting fake news articles and legitimate articles from websites like BuzzFeed. The model implemented in this paper showed a marked improvement over existing models using symbolic and numeric algorithms, data imputation, and Naive Bayes. Having studied about XGBoost, study [14] compares it with other models such as support vector machines (SVMs) and Naïve Bayes on the UA fake news dataset. They also explore different data-preprocessing techniques to improve the performance of these models. Some techniques they worked on are Bag-of-Works (BoW), TF-IDF, and Principal Component Analysis (PCA), thereby improving at least 12% in accuracy and runtime.

**For Informal News Articles (Social Media Posts)** Study [15] focuses on the task of detection of fake news from informal sources, using a unique approach of using a large-scale dataset of tweets with low-quality labels rather the other way: an approach called weak supervision. The authors of the study [15] use various user-level, text-level, and tweet-level features to train different ML models, boasting that such a vast diversity of features were never used before.

### *1.2.2 Deep Learning Approaches*

**For Formal News Articles** Study [16] is a heuristic-driven uncertainty-based ensemble framework for fake news detection in informal news articles and tweets. The approach uses transfer learning for the classification and a soft-voting schema for preprocessing. The model uses a statistical feature fusion network (SFFN) with uncertainty estimation, and a heuristic post-processing technique that considers attributes like the URL, author, source, and username handles as statistical features. The framework is good overall performance and robustness. Two datasets were used: the CONSTRAINT COVID-19 Fake News Detection in English and the FakeNewsNet dataset consisting of political news articles.

A benchmark study conducted in [17] evaluates the performance of 19 machine learning approaches to detect fake news. The datasets used in this study are the LIAR political news dataset, the Fake or Real News dataset focusing on the 2016 US election, and the Combined Corpus dataset consisting of miscellaneous topics. The study concludes that BERT-based methods perform better, and LSTM-based approaches perform well on adequate data size.

**For Informal News Articles (Social Media Posts)** FakeFinder is an app proposed in [18] to detect fake news on live Twitter streams having a comparable accuracy but significantly less response time and memory allocation. Since the model for this app is intended to be used on mobile devices, the focus of FakeFinder is to have a short response time and hence uses a compact language model named ALBERT, a compact version of the state-of-the-art BERT language model. The dataset used in this app was a collection of tweets and comments from Twitter.

Survey [19] introduces a method using Multimodal Variational Autoencoder (MVAE) trained with a Fake News Detector using social media post content, including text and images, to detect fake news. Two real-world datasets used were the Twitter dataset from the 2016 MediaEval benchmark and the Weibo dataset, consisting of Xinhua News Agency and Weibo data. The MVAE significantly improved over the state-of-the-art architectures VQA, Neural Talk, att-RNN, and EANN.

Being acquainted with preceding works, let us introduce the central techniques in our work, their history, and their applications. Coming to a brief history of SVD, it was proposed in 1970 [20], initially applied to the domains of data analysis [6], and recently being used as a standard feature reduction method for various ML and DL methods and outlier detection. The only issue with SVD is its scalability issues for massive datasets, but improved version such as truncated SVD helps address this problem. LSTMs introduced in 1997 [21] are an improvement on RNNs to avoid vanishing-gradient problems. They are great for applications with sequential inputs and are majorly used in the NLP domain. LSTMs are of various kinds and types, such as Bidirectional LSTMs and Stacked LSTMs, but have a common sequential processing and parallelization problem that requires advanced hardware and software solutions.

## 1.3 Proposed Work

### *1.3.1 Formal Definition*

$A$ is the set of news articles, an article being denoted by $a$, and $L$ is the set of corresponding labels, with each label being represented as an $l$. Any article $a$ in set $A$ is a set of words, and labels $l$ in $L$ are either 0 or 1, indicating that the corresponding article is fake or real, respectively. The task is to train an RNN model [22] $M$ using $A$ and $L$ such that when an unseen article $a'$ is given to model $M$, it should predict the correct label for $a'$ as either 0 or 1. After obtaining the set of reduced representations for $A$, we denote here with an $R$. We train the RNN model $M$ using $R$ rather than $A$. The novelty of our method can be briefly stated in the following manner. The fake news detection models in the recent literature can be denoted as $RM$s. If the average training time of all $RM$s is $\hat{t}$ and their average accuracy is $\hat{a}$, then our proposed work has training time $\overline{t}$ and accuracy is $\overline{a}$ where: $\overline{t} < \hat{t}$ and $\overline{a} \approx \hat{a}$.

### *1.3.2 Details*

Our proposed work (Algorithm 1.2) makes training, testing, and deployed usage of such a model $M$ faster. Furthermore, this is achieved by passing the news articles through a feature reduction algorithm in the data preprocessing phase. The feature reduction algorithm we are deploying is called *Truncated SVD*, as described in Algorithm 1.1. When a vectorized representation of an article $v$ passes through *Truncated SVD*, it outputs a reduced representation $r$ of the corresponding vectorized representation $v$ where length($r$) < length($a$). This vectorizer can be a count vectorizer [23], one-hot encoding, or word embedding such as GloVe [24] or word2vec [25], among various others.

Singular-Valued Decomposition (SVD) is a method in linear algebra for the factorization of a matrix. It decomposes a given matrix into three matrices as described in Algorithm 1.1. For a given matrix $A$, initially, the values of $A^T$ and $A^TA$ are computed. Then the eigenvalues of $A^TA$ are calculated, sorted in the descending order of their absolute values, and their square roots are obtained to generate the singular values of $A$. A diagonal matrix $S$ is constructed by placing the singular values of $A$ in descending order over the diagonal. $S^{-1}$ is also calculated. Then, the eigenvectors of $A^TA$ are computed and are placed along with the columns of a matrix called $V$. $V^{-1}$ is also calculated. We then calculate the SVD as $\text{SVD} = \text{AVS}^{-1}$. Here, we take the truncated SVD [26] by taking the $n$ most significant singular values and corresponding left and right singular vectors [27].

**Algorithm 1.1 Truncated SVD**

Inputs:
$Vect(A)$: The matrix for which the truncated SVD is to be computed
$n$: The number of components the data is to be reduced to
Output:
$TruncatedSVD$: The truncated SVD computed for the input matrix
procedure Truncated SVD($Vect(A)$, $n$)
Compute $Vect(A^T)$ and $Vect(A^TA)$
$SingularValues(A) \leftarrow Sort(Eigenvalues(Vect(A^TA)))$
$S \leftarrow DiagonalMatrix\ (Singular(A))$
$S^{-1} \leftarrow Inverse\ (S)$
$V \leftarrow DiagonalMatrix\ (Eigenvectors(Vect(A^TA)))$
$V^T \leftarrow DiagonalMatrix\ (Eigenvectors(Vect(A^TA)))^T$
$TruncatedSVD \leftarrow Truncate(AVS^{-1}, n)$
return $TruncatedSVD$

For a better understanding of the work, let us consider a hypothetical example; there is an article "Trump has won the 2020 US elections," with a label 0, that is, it is a fake news article. When the article is passed through the proposed work, first using SVD, the article is reduced to a representation like [1.238, 4.56, 0.87]. This representation has a length of 3, instead of the article's length, which is 7. This reduced representation is used to train the RNN to learn the label of the article. The above example is illustrated in Fig. 1.2.

**Algorithm 1.2 Proposed Solution**

Inputs:
$X$: The set of articles
$y$: The set of corresponding labels of $A$
$M$: The untrained LSTM model
$n$: The number of components the data is to be reduced to
Outputs:
$M'$: The trained LSTM model
procedure BuildModel($X$,$y$,$M$,$n$)
$Vect(X) \leftarrow$ Preprocessing($X$)
$R \leftarrow TruncatedSVD(Vect(X),n)$
$M' \leftarrow BackPropagation(R,y,M)$ [28]
return $M'$

$A$ is the set of articles $a$, whereas $a$ is a set of words. $L$ is the set of corresponding labels $l$, which is either 0 or 1. $A$ is passed into the Preprocessing function.

**Algorithm 1.3 Preprocessing**

Inputs:

$A$: The set of articles

Outputs:

$A'$: The processed set of articles

1: procedure Preprocessing($A$)

2: $Tokens(A) \leftarrow Tokenizer(A, number\_of\ \_words)$

3. $A' \leftarrow$ *Filter* ($Tokens(A)$, $lower()$, *remove_url*, *punctuation*, *whitespace*, *numeric*, *tags*, *stopwords*, *shortwords*)

4: $Vect(A) \leftarrow CountVectorizer(A', number\_of\_\_words)$

5: return $Vect(A)$

The Preprocessing function (Algorithm 1.3) tokenizes the articles into set of tokens, with the size of the set being the average length of articles in the dataset. The list of all tokens $Token(A)$ is passed into a filter, which filters out those tokens that are URLs; punctuations such as "," "." and ";"; white-spaces; numbers such as 1, 2, and 3.87; XML or HTML tags such as <p>; and stopwords or short words such as "a" and "an." The Preprocessing function at the end returns the count vectorizer of the filtered tokens with each count vectorizer being of the length equal to the average length of the articles in the dataset. Count vectorizer is a conversion technique to make text into numerical data by creating a matrix of the number of occurrences of a given word for a given column and article (Fig. 1.1).

The set of news articles along with their labels is the data used for training the untrained RNN model. The articles are passed into the method Preprocessing (Algorithm 1.3), which returns the processed set of articles. This set along with the parameter "number of components" is passed into the method Truncated SVD (Algorithm 1.1). The truncated set of articles is returned by the method, which along with the set of labels is passed to the method back-propagation, which returns the trained RNN model.

## 1.4 Experimentation

We have selected two fake news datasets: one from Kaggle [29] and the other one named ISOT Fake news detection dataset. The Kaggle dataset contains two comma-separated values (CSV) files, one being "train.csv" and the other being "test.csv". The training datasets had the columns for the news article's title, the text of the article, and the corresponding label that marks an article as either fake or real. The number of articles in both datasets has no considerable class imbalance problem.

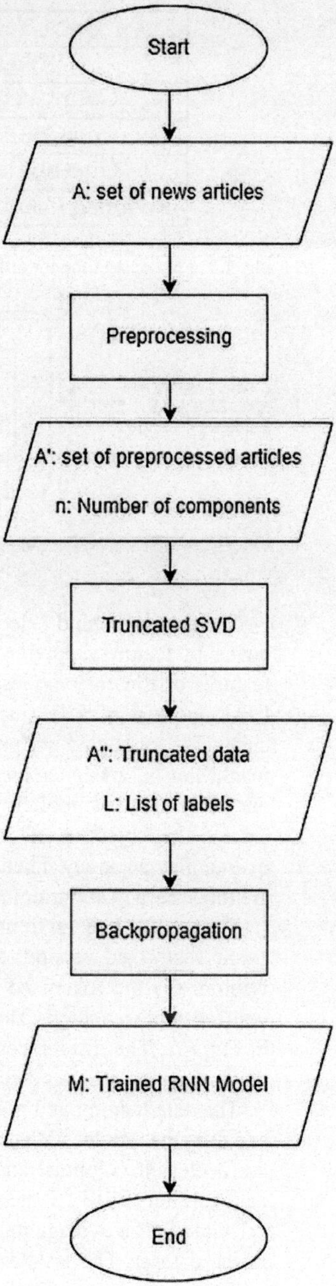


**Fig. 1.1** Flowchart of the proposed work

| Data Version | Processing Time (sec) |
|---|---|
| Tokens | 44.08 |
| Count Vectorizer | 33.83 |
| Reduced Count Vectorizer | 53.83 |
| One-hot Encoder | 43.87 |
| Reduced Count Vectorizer | 47.45 |

**Fig. 1.2** Processing time for different versions of data for experiments

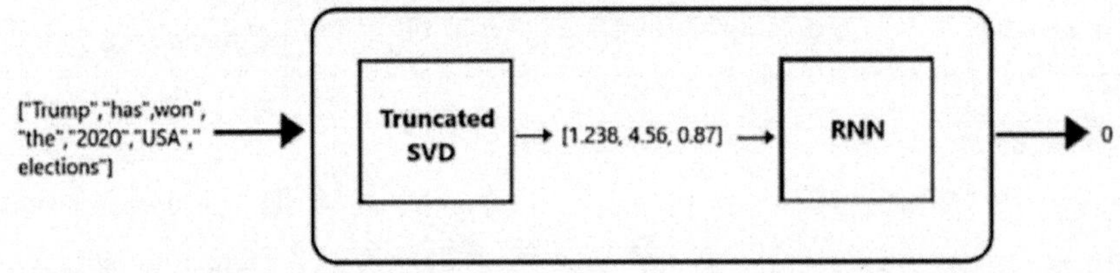


**Fig. 1.3** Overview of the proposed work

The reason behind selecting these datasets is the length of the articles, as the power of feature reduction can only be projected when we have data with more features or dimensions. Various datasets in the literature, such as BuzzFeed and LIAR, consist of a few articles of relatively shorter length, which makes them unsuitable for testing our proposed work. Experiments and results on those datasets would not be a fair testing ground for our model. Also, the dataset we used for our experiments is publicly available and did not have any data privacy or integrity issues. The objective of our research work is to reduce the training time while maintaining accuracy. Therefore, the metrics of paramount importance to this study are the accuracy and training time.

There were five versions of data that we are using for experimentation: (1) the count vectorized version, (2) the reduced representation of the count vectorized version, (3) the tokenized version, (4) the one-hot encoded version, and (5) its reduced representation. The processing time to obtain each version can be found in Fig. 1.2. The differences between these data used for training different models are due to the data specification required for each model (Fig. 1.3).

The data loading and processing steps are simple. We read the data from the file and drop the articles with shorter lengths. The selected articles are filtered to remove the irrelevant columns, and their informative columns are merged into a single-data column called "Sentences." The average length of these remaining articles is calculated. The average helps construct a tokenizer of suitable size for the articles in the dataset. The articles shorter than the average length are padded. Thereby, we obtain the representation that we call the tokenized version, which is used to train the LSTM, RNN, and GRU models. The next version is formed by passing the tokenized data into a count vectorizer object to form a count vectorized version.

This version is used to train the models of SVM and Random Forest. The reduced representation of the count vectorized version is formed by passing it to a Truncated SVD algorithm that transforms the input data and selects the top 100 vectors. This version is used to train all the models (with SVD). The next version is formed by passing the tokenized data into a one-hot encoder object to form a one-hot encoded version. This version is used to train the models of SVM and Random Forest. The reduced representation of the one-hot encoded version is formed by passing the one-hot encoded version into a Truncated SVD algorithm, again selecting the top 100 vectors. This version is used to train all the models (with SVD).

## 1.5 Results and Discussion

Our experiments not only tried comparing our proposed work with other prominent works such as [5, 13, 14] as discussed in the literature but also altered versions of those works. The reason behind creating shallow RNN, LSTM, and GRU models is that we are trying to establish the novelty of our method, which could have been overshadowed if we used deeper networks, as they generally perform well. The other reason is that it takes very long to train the model with a single layer; adding more layers was practically unreasonable.

The results are presented in Tables 1.1 and 1.2. According to Tables 1.1 and 1.2, our proposed model has relatively high accuracy with one of the lowest training times. The proposed model has a better accuracy versus training time trade-off than all the other models in our experiments. The entries in Tables 1.1 and 1.2 that have an asterisk (*) were ran only a single time and not averaged over three rounds of training unlike the rest of the entries. The entries with "–" in Table 1.2 consumed a lot of space and took a lot of time to train, making them inappropriate for a fair comparison.

## 1.6 Conclusion and Future Scope

The authors conclude through their study that reducing the features has a huge impact on the training and testing time. Deploying DL models such as RNN to NLP tasks such as fake news detection results in exceptional performance even if most of the features from the dataset were removed. In our future works, we plan to experiment with different vectorizers and different feature reduction algorithms apart from SVD and applying the work to different NLP tasks such as text summarization, sentiment analysis, and topic modeling. Precision and recall scores for the experiments conducted shall also aid in establishing the novelty of the method, which shall be extensively studied and presented in future studies.

**Table 1.1** Results on Kaggle fake news dataset

| Model name | Vectorizer | With/without SVD | Accuracy (%) | Training time (s) |
|---|---|---|---|---|
| SVM | Count Vectorizer | With | 88.41 | 11.09 |
| | One-Hot Encoder | 63.21 | 41.10 | |
| | Count Vectorizer | Without | 92.36 | 142.81 |
| | One-Hot Encoder | 64.27 | 85.45 | |
| Random Forests | Count Vectorizer | With | 85.43 | 8.48 |
| | One-Hot Encoder | 63.83 | 28.41 | |
| | Count Vectorizer | Without | 91.14 | 44.92 |
| | One-Hot Encoder | 67.93 | 19.28 | |
| LSTM | Count Vectorizer | With | 88.21 | 3.72 |
| | One-Hot Encoder | 61.9 | 19.34 | |
| | Count Vectorizer | Without | 72.12 | 28800.34* |
| | One-Hot Encoder | 61.93 | 7.32 | |
| GRU | Count Vectorizer | With | 87.10 | 3.30 |
| | One-Hot Encoder | 63.53 | 6.78 | |
| | Count Vectorizer | Without | 73.61 | 16552.83* |
| | One-Hot Encoder | 60.72 | 10.63 | |
| Proposed Work | Count Vectorizer | With | 87.00 | 1.70 |
| RNN | One-Hot Encoder | 61.66 | 3.95 | |
| | Count Vectorizer | Without | 70.20 | 17897.42* |
| | One-Hot Encoder | 60.72 | 10.63 | |

**Table 1.2** Results on ISOT fake news dataset

| Model name | Vectorizer | With/without SVD | Accuracy (%) | Training time (s) |
|---|---|---|---|---|
| SVM | Count Vectorizer | With | 99.06 | 26.08 |
| | One-Hot Encoder | 75.57 | 95.80 | |
| | Count Vectorizer | Without | 99.64 | 307.87 |
| | One-Hot Encoder | 78.20 | 390.23 | |
| Random Forests | Count Vectorizer | With | 96 | 17.23 |
| | One-Hot Encoder | 75.17 | 25 | |
| | Count Vectorizer | Without | 99.81 | 95.83 |
| | One-Hot Encoder | 91.94 | 27.88 | |
| LSTM | Count Vectorizer | With | 99.20 | 29 |
| | One-Hot Encoder | 57.75 | 4.40 | |
| | Count Vectorizer | Without | 69.96 | 7174* |
| | One-Hot Encoder | – | – | |
| GRU | Count Vectorizer | With | 99.20 | 20.08 |
| | One-Hot Encoder | 63.60 | 15.90 | |
| | Count Vectorizer | Without | 74.87 | 6831* |
| | One-Hot Encoder | – | – | |
| Proposed Work | Count Vectorizer | With | 98.90 | 17.70 |
| RNN | One-Hot Encoder | 61.80 | 2.70 | |
| | Count Vectorizer | Without | 73.22 | 5702* |
| | One-Hot Encoder | – | – | |

## References


1. A. Bondielli, F. Marcelloni, A survey on fake news and rumour detection techniques. Inf. Sci. **497**(2019), 38–55 (2019)
2. X. Che, D. Metaxa-Kakavouli, J.T. Hancock, Fake news in the news: An analysis of partisan coverage of the fake news phenomenon, in *Companion of the 2018 ACM Conference on Computer Supported Cooperative Work and Social Computing*, (ACM, 2018), pp. 289–292
3. X. Zhou, R. Zafarani, A survey of fake news: Fundamental theories, detection methods, and opportunities. ACM Comput. Surv. **53**(5), 1–40 (2020)
4. M.F. Mridha, A.J. Keya, M.A. Hamid, M.M. Monowar, M.S. Rahman, A comprehensive review on fake news detection with deep learning. IEEE Access **9**, 156151–156170 (2021)
5. S.A. Alameri, M. Mohd, Comparison of fake news detection using machine learning and deep learning techniques, in *2021 3rd International Cyber Resilience Conference (CRC)*, (IEEE, 2021), pp. 1–6
6. C.O.S. Sorzano, J. Vargas, A.P. Montano, A survey of dimensionality reduction techniques. arXiv preprint arXiv:1403.2877 (2014)
7. K.M. Yazdi, A.M. Yazdi, S. Khodayi, J. Hou, W. Zhou, S. Saedy, Improving fake news detection using k-means and support vector machine approaches. Int. J. Electron. Commun. Eng. **14**(2), 38–42 (2020)
8. J.I.Z. Chen, J.I. Zong, Automatic vehicle license plate detection using k-means clustering algorithm and CNN. J. Electr. Eng. Autom. **3**(1), 15–23 (2021)
9. L. Nanni, S. Ghidoni, S. Brahnam, Handcrafted vs. nonhandcrafted features for computer vision classification. Pattern Recogn. **71**(2017), 158–172 (2017)
10. L. Shi, J. Zhang, E. Liu, P. He, Text classification based on nonlinear dimensionality reduction techniques and support vector machines, in *Third International Conference on Natural Computation (ICNC 2007)*, vol. 1, (IEEE, 2007), pp. 674–677
11. S.R. Sekhar, G.M. Siddesh, M. Raj, S.S. Manvi, Protein class prediction based on Count Vectorizer and long short term memory. Int. J. Inf. Technol. **13**(1), 341–348 (2021)
12. J.P. Haumahu, S.D.H. Permana, Y. Yaddarabullah, Fake news classification for Indonesian news using Extreme Gradient Boosting (XGBoost). IOP Conf. Ser.: Mater. Sci. Eng. **1098**(5), 052081 (2021) IOP Publishing
13. J. Antony Vijay, H. Anwar Basha, J. Arun Nehru, A dynamic approach for detecting the fake news using random forest classifier and NLP, in *Computational Methods and Data Engineering*, (Springer, 2021), pp. 331–341
14. H. Wang, Y. Ma, Y. Deng, Y. Wang, Fake news detection algorithms comparison and application of XGBoost, SVM, and NB. World Sci. Res. J. **7**(1), 323–329 (2021)
15. S. Helmstetter, H. Paulheim, Collecting a large scale dataset for classifying fake news tweets using weak supervision. Future Internet **13**(5), 114 (2021)
16. S.D. Das, A. Basak, S. Dutta, A heuristic-driven uncertainty based ensemble framework for fake news detection in tweets and news articles. Neurocomputing **491**, 607–620 (2022)
17. J.Y. Khan, M.T.I. Khondaker, S. Afroz, G. Uddin, A. Iqbal, A benchmark study of machine learning models for online fake news detection. Mach. Learn. Appl. **4**(2021), 100032 (2021)
18. L. Tian, X. Zhang, M. Peng, FakeFinder: Twitter fake news detection on mobile, in *Companion Proceedings of the Web Conference 2020*, (ACM, 2020), pp. 79–80
19. D. Khattar, J.S. Goud, M. Gupta, V. Varma, MVAE: Multimodal variational autoencoder for fake news detection, in *The World Wide Web Conference*, (ACM, 2019), pp. 2915–2921
20. J. Zhang, P. Zhang, X. Bin, Analysis of college students' public opinion based on machine learning and evolutionary algorithm. Complexity **2019**, 1712569 (2019)
21. K. Smagulova, A.P. James, A survey on LSTM memristive neural network architectures and applications. Eur. Phys. J. Spec. Top. **228**(10), 2313–2324 (2019)
22. S. Hochreiter, J. Schmidhuber, Long short-term memory. Neural Comput. **9**(8), 1735–1780 (1997)

23. G. Eason, B. Noble, I.N. Sneddon, On certain integrals of Lipschitz-Hankel type involving products of Bessel functions. Phil. Trans. R. Soc. London **A247**, 529–551 (1955)
24. J. Pennington, R. Socher, C.D. Manning, Glove: Global vectors for word representation, in *Proceedings of the 2014 Conference on Empirical Methods in Natural Language Processing (EMNLP)* (2014), pp. 1532–1543
25. Y. Goldberg, O. Levy, word2vec explained: Deriving Mikolov et al.'s negative-sampling word-embedding method. arXiv preprint arXiv:1402.3722 (2014)
26. N. Halko, P.-G. Martinsson, J.A. Tropp, Finding structure with randomness: Stochastic algorithms for constructing approximate matrix decompositions. arXiv preprint arXiv:0909.4061 (2009)
27. E. Garcia, *Singular Value Decomposition (SVD) – A Fast Track Tutorial. Using the Singular Value Decomposition* (E. Garcia, 2006)
28. M.C. Mozer, A focused backpropagation algorithm for temporal, in *Backpropagation: Theory, Architectures, and Applications*, (ACM, 1995), pp. 137–169
29. Kaggle Fake News Dataset. Retrieved May 17, 2022, from https://www.kaggle.com/competitions/fake-news/data

# Chapter 2
# Using Convolutional Neural Network to Enhance Coronary Heart Disease Predictions in South African Men Living in the Western Cape Region

**Elias Tabane**

## 2.1 Introduction

During this age of technology and digitalization, data have proven to be the fuel of organizations and industries. The healthcare industry isn't far behind in this respect. Nowadays, most hospitals and medical institutes have their patients' data stored in electronic formats [1]. This includes their medical histories, their symptoms, their diagnoses, the durations of their illnesses, their recurrences, and any fatalities. As a result, the number of medical data generated every day is consistently increasing. However, this wealth of data is typically left untapped because of a lack of effective analytical tools, methods, and personnel to draw insights and hidden relationships from this data. If the data at hand are used to develop screening and diagnostic models, they will not only reduce the strain on medical personnel but also aid in early detection and prompt treatment for patients, thereby drastically improving the health system [2, 3].

In recent years, researchers and experts working within the medical field have started realizing the immense amount of information available in these medical data sets, thereby inspiring medical analyses of data for diseases such as dementia, Alzheimer, tuberculosis, diabetes, cancer, etc. Amid this vast array, one of the predominant and most significant diagnoses within the sector of health analysis is coronary heart disease (CHD). Coronary arteries play crucial roles in delivering oxygen to the central muscle. According to the Southern Cross Medical Care Society of New Zealand, the constant buildup of fat or bad cholesterol within these

E. Tabane (✉)
University of South Africa, Pretoria, South Africa
e-mail: Tabane@unisa.ac.za

 
I. Woungang, S. K. Dhurandher (eds.), *5th International Conference on Wireless, Intelligent and Distributed Environment for Communication*, Lecture Notes on Data Engineering and Communications Technologies 174,
https://doi.org/10.1007/978-3-031-33242-5_2

arteries' walls leads to their narrowing down and to eventual blockage, thereby giving rise to CHD [4].

A mid-level blockage might just cause initial discomfort and alterations in the person's lifestyle. However, when the flow of oxygen through the coronary arteries is severely hampered, it can become fatal. The danger factors associated with CHD are often a mix of controllable factors like those influenced by one's lifestyle and uncontrollable factors like age, ethnicity, family medical history, etc. [5]. An early detection of CHD symptoms can help the patient to manage a variety of those risk factors through lifestyle changes and/or medication, thus preventing this disease from developing into a severe form that would prove fatal [6]. Machine learning (ML) plays a big role in disease prediction [7]. It predicts whether the patient features a disease type or not, supported by an efficient learning technique [8].

In this chapter, we are utilizing supervised learning techniques for predicting the first stage of a heart condition. Ensemble algorithms and a number of other algorithms, like a k-nearest neighbor (KNN), support vector machine (SVM), decision tree (DT), naïve Bayes (NB), and random forest (RF), classify whether the people tested belong to the class of having a heart condition or the class of having a healthy heart, and we evaluate the accuracy of these algorithms against the convolutional neural network algorithm to determine which one can accurately predict heart disease.

The rest of this chapter is structured as follows: Section 2.2 details the literature review of the current research in this field. Section 2.3 describes the proposed architecture and methodology. In Section 2.4, the experimental results and a comparison between classification techniques are presented. Finally, Section 2.5 concludes the chapter.

## 2.2 Literature Review

The research work of Sharma, et al. [9] established the use of an artificial neural network (ANN) and an RSM through back-propagation-algorithm-based mathematical modeling. The experiments performed the machining characteristic optimization of a micro-EDM during the microhole machining operation on Ti–6Al–4V. The input parameters were applied for the development of the ANN predicting model.

Gopal [10] fostered an ensemble slope-boosting choice tree model for the assessment of client enthusiasm and its connection with search ads for two kinds of features, namely dynamic and static features. A Bing Ads data set was utilized to assess the created strategy. The outcomes were profoundly encouraging and could effectively adapt to the connected errand with high supportability. Li [8] proposed classification methods that were supported by a nonlinear SVM with polynomial and Gaussian kernels and by output coding (OC). Ensembles of learning machines separate normal tissues from malignant ones, classify different types of lymphoma, and research the role of sets of coordinately expressed genes in carcinogenic processes of lymphoid tissues. By using phenomenon data from "Lymphochip", the author of [8] has shown that SVM can correctly separate tumoral tissues and that OC ensembles are often successfully used to classify different kinds of lymphoma.

Lin et al. [11] proposed a support-vector-based fuzzy neural network (SVFNN) to attenuate the training and testing errors, to improve performance. They developed a learning algorithm consisting of three learning phases in order to construct the SVFNN, during which the fuzzy rules and membership functions are automatically determined by the clustering principle. To research the effectiveness of the proposed SVFNN classification, they applied the corresponding model to various data sets from the University of California–Irvine (UCI) Repository and Statlog collection. The experimental results show that using the proposed SVFNN for pattern classification can achieve good classification performance with a drastically reduced number of fuzzy kernel functions. In their work, [12] proposed a self-advising SVM method for the improvement of SVM performance by transferring more information from the training phase to the testing phase. This information is generated by using misclassified data during the training phase. Their study showed an improvement in accuracy, and the F-score and statistical tests reveal the importance of these improvements. They claimed that by using the misclassified data during the training phase, overtraining is often avoided in their proposed method.

The convolutional recurrent neural network (CRNN) model was proposed by in [13] to differentiate and characterize electrocardiogram (ECG) signals for detecting arrhythmias. CRNN is a blend of CNN and RNN; however, this calculation doesn't utilize the completely associated layer, however useful it is for obtaining a general perspective. The model learns directly from the source, so it doesn't require point-by-point explanations of the sample information, and the length of the sample information isn't restricted. The preprocessing task involves standardization. All in all, the result of the general exhibition of the network was 98.81% successful. The defect recognized here was connected to a deficiency in the ECG data set, so it was difficult to determine whether the calculation accurately detected heart infections.

One more technique considered for characterizing arrhythmias was utilizing something similar and unique: patients' ECG recordings utilizing a convolutional neural network. The technique used in this chapter differed, though, by dividing the ECG information into two sections. The initial segment was for the intrapatient worldview. In this part, the preparation and the test information were taken from similar patients, which took a one-sided approach in the neural network. The subsequent part was for the interpatient worldview. The preparation and the test information come from various patients. For characterization, a multifaceted perceptron was presented as against the completely associated layer in CNN. However, the yield result was not tremendous [14].

Sajja et al. [15] proposed a CNN technique to predict heart disease at an early stage. Their focuses were specifically on a comparison between the traditional approaches, such as logistic regression, k-nearest neighbors (KNN), naïve Bayes (NB), support vector machine (SVM), neural networks (NNs), and the proposed prediction model of a CNN. The UCI machine-learning repository data set for experimentation and cardiovascular disease (CVD) predicts with 94% accuracy.

The work of [16] involves the application of a convolutional neural network for predicting the occurrence of ventricular tachyarrhythmia. Using heart rate variability features in their exploration work, the authors used a one-dimensional convolutional neural network (1D CNN) to remove highlights from pulse changeability (HRV: heart rate variability) and, accordingly, to predict the beginning of VTA. They likewise compared the prediction execution of our CNN with other machine-inclining (ML) calculations, like an artificial neural network (ANN), a support vector machine (SVM), and a k-nearest neighbor (KNN), which used 11 HRV highlights that were extracted by utilizing traditional methods.

The proposed CNN achieved generally higher prediction exactness rates of 84.6% than the ANN, SVM, and KNN calculations, which obtained prediction rates of 73.5%, 67.9%, and 65.9%, respectively, by utilizing 11 HRV highlights. Their outcomes showed that the proposed 1D CNN could further develop VTA prediction precision by coordinating the data cleaning, preprocessing, and highlighting extractions. Furthermore, prediction was centered on deep convolutional neural networks, which predicted cardiovascular risk from computed tomography. They developed a deep-learning system to automatically identify individuals who were at high risk for cardiovascular disease and tested the system performance in four large independent holdout cohorts with a variety of clinical presentations and computed tomography (CT) scanning techniques.

## 2.3 The Proposed Heart Disease Prediction System

The main objective of the proposed system technique is to use convolutional neural networks and other techniques to enhance the performance of prediction. Figure 2.1 unpacks the architecture of the proposed system. It's structured into six stages: data collection, data preprocessing, feature selection, data splitting, training models, and evaluating the models.

### *2.3.1 Data Collection*

Th data set for heart disease prediction, which is used for both training and the evaluations, consists of 406 records and features and one target column, which represents either class 1, indicating the presence of heart disease, or class 0, which indicates the absence of heart disease. Table 2.1 describes all these features.

### *2.3.2 Data Selection*

The selected features have been scaled to appear within the ranges of {0,1]}; furthermore, all the values that were detected as missing have been removed.

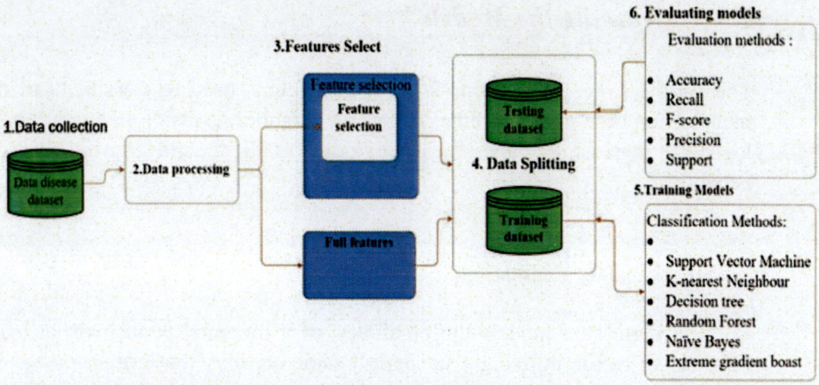


Fig. 2.1 The structure of the proposed systems

**Table 2.1** Heart disease data set descriptions

| Feature | Description |
|---|---|
| SBP | Systolic blood pressure |
| Tobacco | Cumulative tobacco (kg) |
| LDL | Low-density lipoprotein cholesterol (HDL cholesterol levels greater than 60 milligrams per deciliter (mg/dl) are high) |
| Adiposity | Adiposity is associated with several secondary diseases (sequalae), like (type 2) diabetics, high blood pressure, cardiovascular disease, fatty liver, and disorders of the adipose, the fat distribution of the body determines the risk of developing these secondary diseases |
| Famihist | Family history of the heart disease (present = 1, absent = 0) |
| Typea | The state of being grossly overweight |
| Alcohol | Current alcohol consumption |
| Age | Age at onset |
| CHD | Coronary heart disease (yes = 1 or no = 0) |

### *2.3.3 Feature Selection*

The extraction of the best features may be a crucial phase because irrelevant features often affect the classification efficiency of the machine-learning classifier.

### *2.3.4 Data Splitting*

In this step, the data set for heart disease predictions is divided such that 70% is for the training set and 30% is for the testing set. The training set is used to train the models, and the testing set is utilized to evaluate the models. Also, a tenfold cross validation is utilized in the training set.

## 2.3.5 *Training the Models*

The various types of machine-learning algorithms used to classify heart disease include support vector machine, K-nearest neighbor, decision tree, random forest, logistical regression, naïve Bayes, extreme gradient, and convolutional neural networks.

### 2.3.5.1 Logistic Regression

Logistic regression models the prediction of a binomial result with at least one informative factor, utilizing a standard logistic capacity that estimates the connection between the absolute dependent variable and at least one independent factor by assessing probabilities. The logistic capacity is given by $f(x) = 11 + e - x$, which is known as the sigmoid bend. SVM is a paired arrangement calculation that creates an $(n - 1)$-dimensional hyperplane to isolate two classes in an $n$-dimensional plane. The characterization hyperplane is constructed in a high-dimensional space that addresses the biggest division, or edge, between the two classes.

### 2.3.5.2 Random Forest

Random forests are outfit learning calculations where decision trees are averaged and trained on different pieces of the preparation set to reduce differences and avoid overfitting. Calculating random forests utilizes sacking or bootstrap accumulating, and at each split, a random subset of provisions is selected. Sacking is an equal outfit because each model is independently constructed. Boosting contains a successive group where each model is constructed on the basis of remedying the misclassifications of the most recent model. In boosting methods, the loads are initialized on preparing tests, and for $n$ cycles, a classifier is trained by utilizing a solitary element and its preparing blunders are evaluated. At that point, the classifier with the fewest blunders is selected, and the loads are updated; accordingly, the last classifier is formed as a blend of $n$ classifiers. A lift classifier is in the structure $FT(x) = \sigma ft(x)Tt = 1$, where each $ft$ is a weak student and $x$ is their information. Each weak student produces a yield speculation, $h(xi)$, for each example in the preparation set. At every emphasis $t$, a weak student is selected and assigned a coefficient $\alpha t$, where the end goal is that the aggregate of preparing mistakes $Et$ of the subsequent $t$-stage help classifier is minimized.

### 2.3.5.3 K-Nearest Neighbor

The k-nearest neighbor (KNN) could also be a nonparametric classification algorithm; i.e., it doesn't make any presumptions on the elementary data set. It's known for its simplicity and effectiveness. It is a supervised learning algorithm. A labeled training data set is provided, and here the data points are categorized into various classes so that a class of the unlabeled data is often predicted. When some new data are added, the KNN classifies the data accordingly. It's more useful during a data set that is roughly divided into clusters and belongs to a specific region of the data plot. Thus, this algorithm more accurately and more clearly divides the data inputs into different classes. KNN determines the category that has most points and has the least distance from the data point that must be classified. Hence, the Euclidean distance must be calculated between the test sample and the specified training samples. After we gather the k-nearest neighbor data, we use the bulk of them to predict the category of the training example.

#### 2.3.5.4 Support Vector Machine (SVM)

SVM is another artificial intelligence (AI) strategy that depends on the machine-learning hypothesis, and it is one of the computational methods created by Vapnik [17]. Given the underlying danger of structural risk minimization (SRM), SVM can obtain dynamic principles, while suffering few blunders, for the autonomous test set and henceforth can effectively resolve the learning issues. Lately, SVM is applied to resolve nonlinear, nearby least, and high measurement issues. In numerous commonsense applications, SVM can guarantee higher precision for a drawn-out expectation compared with that of other computational methods. SVM depends on the idea of choice planes that characterize choice limits. SVM makes a hyperplane by utilizing a straight model to carry out nonlinear class limits through some nonlinear planning input vectors and into a high-dimensional element space [10]. In SVM, there is some obscure and nonlinear reliance, such as in the planning of capacity for $n = ()$, which is between some high-dimensional information vector $x$ and the scalar yield (or the vector yield $y$ because of multiclass SVM). There are no data on the fundamental joint likelihood capacities, and only appropriation-free learning should be used.

#### 2.3.5.5 Convolutional Neural Network

Convolutional neural organizations have a neural relationship with one convolutional layer, which is used fundamentally for data with grid-like geologies [18]. Models are 2D pictures of data taking into consideration the characterizations of 1D time-series data (tests at time stretches) with the assistance of a type of numerical activity known as convolution. Heaton [18] defined *CNNs* as "neural affiliations that utilize convolution instead of general design extension in some place close to one of their layers." A CNN was arranged explicitly for the picture assertion issue on PC. The fundamental work on present-day CNNs was introduced by [19] for handwriting attestation, but around 2012, AlexNet achieved unrivaled pictures in the ImageNet challenge [20].

Figure 2.1 represents how the proposed system of how the machine-learning algorithm will be evaluated.

### 2.3.6 *Assessing Models*

An assessment of the proposed model is performed by zeroing in on certain features of the models, specifically their exactness, review, accuracy, F-score, receiver operating characteristic (ROC), and area under the curve (AUC). Exactness is one of the main exhibition measurements. It is characterized by the difference between the right order and the original order, as follows:

$$\text{Accuracy} = (\text{TP} + \text{TN}) / (\text{TP} + \text{TN} + \text{FP} + \text{FN})$$

$$\text{Specificity} = \text{TN}/ (\text{TN} + \text{FP})$$

$$\text{Sensitivity} = \text{TP}/ (\text{TP} + \text{FN})$$

where:

TP = true positive: the number of samples classified as having heart disease when they had it.
TN = true negative: the number of samples classified as not having heart disease when they did not have it.
FN = false negative: the number of samples classified as not having heart disease when they had it.
FP = false positive: the number of samples classified as having heart disease when they did not have it.

$$\text{Ac} = \frac{(\text{TP} + \text{TN})}{(\text{TP} + \text{TN} + \text{FP} + \text{FN})} \tag{2.1}$$

$$\text{Recall} = \frac{\text{TP}}{\text{TP} + \text{FN}} \tag{2.2}$$

Precision is identified as follows:

$$\text{Precision} = \frac{\text{TP}}{\text{TP} + \text{FP}} \tag{2.3}$$

The F-measure is often referred to as the F1-score, and it measures the mean value of precision as follows:

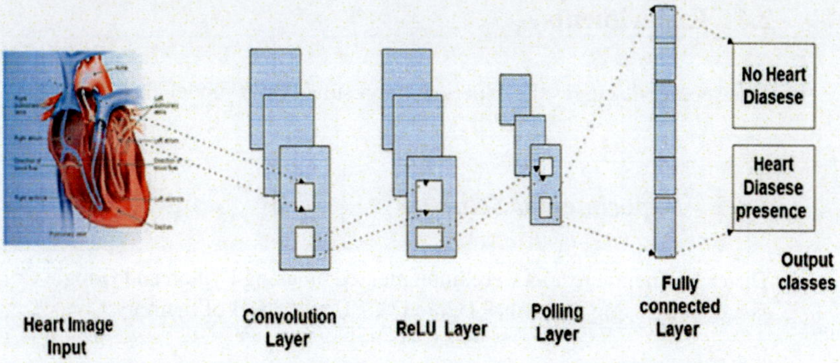


**Fig. 2.2** Proposed convolutional neural network architecture. (Source: Dutta et al. (2020))

$$\text{F-measure} = \frac{(2 \times \text{precision} \times \text{recall})}{(\text{Precision} + \text{recall})} \tag{2.4}$$

The beneficiary working trademark bend (ROC) is a diagram showing the effectiveness of an order calculation including all grouping edges. Two boundaries are displayed in this bend: genuine positive and bogus positive. The region under the bend (AUC) marks a classifier's capacity to separate the classes and is used as a depiction of the ROC bend. The better the AUC is, the higher the model's productivity is in separating between the positive and negative gatherings (Fig. 2.2).

Along these lines, in this proposed design, our coronary illness data set will be pass into the information layer, which will at that point drive the data into their convolutional layer, whose job in the convolution layer is to separate the components from a picture. It is a numerical activity that takes two data sources, such as a picture network and a channel or kernel.

In the subsequent stage, the data have passed into the rectified linear unit (ReLU) layer, which deals with safeguarding the outstanding development in the calculation required to operate the neural network. On the off chance that the CNN scales in size, the computational expense of adding extra ReLUs directly increases. A ReLU utilizes actuation work to determines whether a neuron should be activated, by ascertaining the weighted total and adding inclination to it. The initial work introduces nonlinearity into the yield of a neuron.

The data from a ReLU is then passed on to the pooling layer, whose sole purpose is to pool layers, which are used to reduce the dimensions of the component maps. Subsequently, it reduces the quantity of boundaries to learn and the calculation performed in the network, making the model more powerful to the various elements in the picture. The completely connected layer at that point addresses those layers where every contribution from one layer is connected to each initiation unit of the following layer. Finally, the yield classes reveal whether each data set contains any coronary illnesses.

## 2.4 Experiments

This section discusses the experimental results of our model classification.

### *2.4.1 Experimental Setup*

The experimental results were implemented by using Python and Google Colab and executed by using an Intel (R) Core i7 CPU and 8 GB of memory.

#### 2.4.1.1 Data Set

The data set includes a retrospective sample of men in a region of the Western Cape, South Africa, where people are at high risk for developing heart disease. There are roughly two controls per case of CHD. Many of the CHD-positive men have undergone blood pressure reduction treatment and other programs to reduce their risk factors after their CHD event. In some cases, the measurements were made after these treatments. These data were first taken from a larger data set, described in Rousseauw et al., 1983, in the *South African Medical Journal* [4]. Now the data set is in the public domain on Kaggle.

Data source: SAheart | Kaggle https://creativecommons.org/publicdomain/zero/1.0/

#### 2.4.1.2 Data Type

Table 2.2 describes the data type of the data set attributes.

Figure 2.3 depicts the percentage outcomes of heart disease predictions.

Figure 2.4 depicts the presence or absence of heart disease, which can be traced from family histories.

**Table 2.2** Data type

| Name | Data type |
|---|---|
| SBP | Int64 |
| Tobacco | Object |
| LDL | Object |
| Adiposity | Object |
| Typea | Int64 |
| Obesity | Object |
| Alcohol | Object |
| Age | Int64 |
| Dtype | Object |

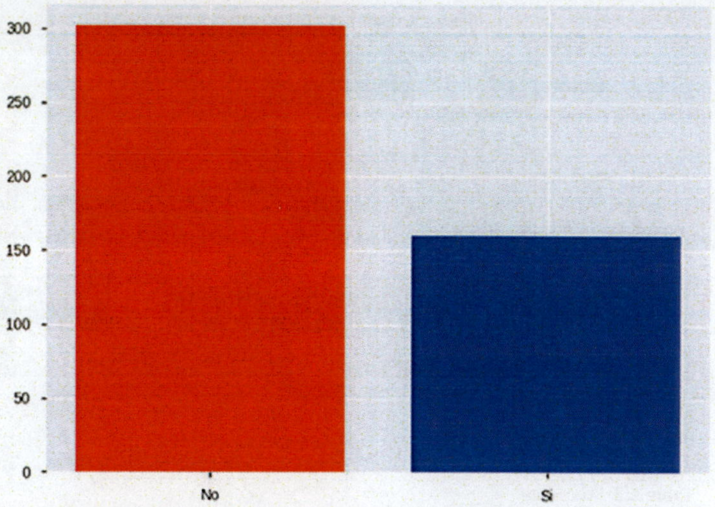


**Fig. 2.3** Heart predictions

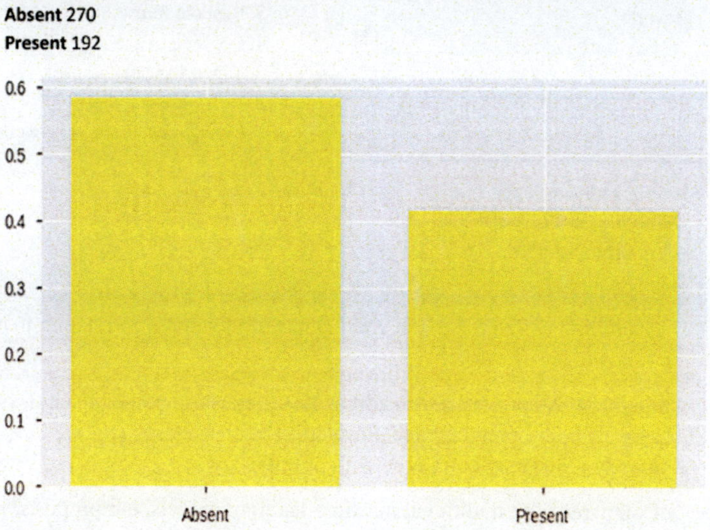


**Fig. 2.4** Family history of heart disease

The histogram suggests that 41% of heart disease cases can be traced back to family history, and 60% of cases have no link to heart disease coming from their family histories.

Figure 2.5 identifies correlations between several variables.

There are some correlations between several numeric features:

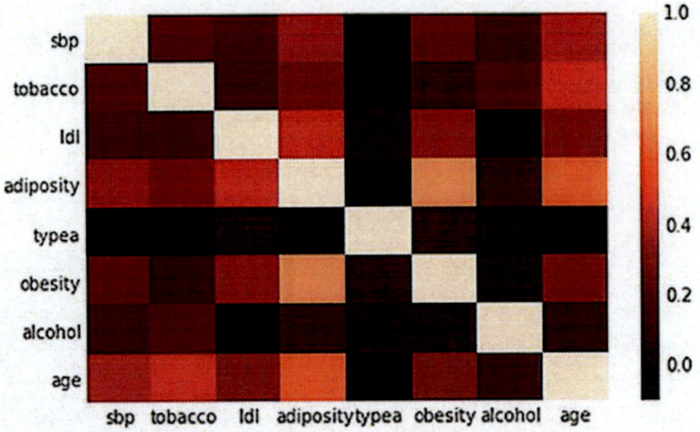


**Fig. 2.5** Identification of correlations between variables

**Table 2.3** Accuracy prediction results for machine-learning algorithms against convolutional neural networks

| Model | Accuracy |
|---|---|
| 1. Logistic regression | 95.83 |
| 2. Naïve Bayes | 81.11 |
| 3. Random forest | 89.72 |
| 4. Decision tree | 79.72 |
| 5. Support vector | 99.16 |
| 6. K-nearest neighbor | 97.70 |
| 7. Convolutional neural network | 99.86 |

Adiposity and obesity
Adiposity and age
Tobacco and age

According to Table 2.3, the coefficients of the above correlation matrix suggests the following:

Age is strongly correlated with adiposity, the consumption of alcohol, smoking, high systolic blood pressure, and elevated LDL cholesterol.
Adiposity is strongly correlated with obesity.

Compared with traditional machine-learning models, our proposed model obtain an accuracy value of 99.86%, which is slightly higher than that of the SVM classifier, which managed to reach 99.16% accuracy (Table 2.4).

When comparing the convolutional neural network with the five-ensemble algorithm, extreme gradient boost did score well, with an accuracy value of 94.72%, but the convolutional neural network did outscore it by reaching 99.86% accuracy.

These outcomes affirm that the CNN classifier beats all the current normally used ML models for coronary illness prediction on precision, for both CHD and non-CHD classes.

**Table 2.4** Accuracy predication for ML models

| Model | Accuracy |
|---|---|
| 1. Bagging classifier | 91.66 |
| 2. AdaBoost classifier | 43.33 |
| 3. Extreme gradient boost | 94.72 |
| 4. Gradient boosting classifier | 89.16 |
| 5. Xgboost | 88.33 |
| 6. Convolutional neural network | 99.86 |

## 2.5 Conclusion

In this chapter, we developed the proposed system to predict heart disease. A convolutional neural network with feature extraction algorithms was used to improve prediction performance for heart disease. The feature extraction algorithms were used to extract essential features from the South African heart disease data set. A comparison between convolutional neural networks and six classifiers (logistical regression, naïve Bayes, random forest, decision tree, support vector, and k-nearest neighbor) was applied to the selected features, and the ensemble models were tested against our proposed model. The experimental results showed that the convolutional neural network algorithm using the feature extraction method achieved the best performance.

## References

1. B.B. Pradhan, B. Bhattacharyya, Modelling of micro-electro discharge machining during machining of titanium alloy Ti–6Al–4V using response surface methodology and artificial neural network algorithm. J. Eng. Manuf. **223**, 683–693 (2009)
2. G. Krishna Mohana Rao, G. Rangajanardhaa, Development of hybrid model and optimization of surface roughness in electric discharge machining using artificial neural networks and genetic algorithm. J. Mater. Process. Technol. **209**, 1512–1520 (2009)
3. R. Atefi, A. Razmavar, F. Teimoori, F. Teimoori, The influence of EDM parameters in finishing stage on MRR of hot worked steel using ANN. J Basic Appl. Sci. Res. **2**(3), 2307–2311 (2011)
4. J. Rousseauw, J. du Plessis, A. Benade, P. Jordaan, J. Kotze, J. Ferreira, Coronary risk factor screening in three rural communities. S. Afr. Med. J. **64**, 430–436 (1983)
5. Q. Gao, Q.-h. Zhang, S. Shupeng, J.-h. Zhang, Parameter optimization model in electrical discharge machining process. J. Zhejiang Univ. Sci. **9**(1), 104–108 (2008)
6. K. Wang, H.L. Gelgele, Y. Wang, Q. Yuan, M. Fang, A hybrid intelligent method for modelling the EDM process. Int J Mach Tool Manu **43**, 995–999 (2003)
7. K.P. Somashekhar, N. Ramachandran, J. Mathew, Optimization of material removal rate in micro-EDM using artificial neural network and genetic algorithms. Mater. Manuf. Process. **25**, 467–475 (2010)
8. E.Y. Li, Artificial neural networks and their business applications. Inf. Manag. **27**, 303–313 (1994) FLEX Chip Signal Processor (MC68175/D), Motorola, 1996 worked steel using ANN. J. Basic Appl. Sci.
9. V. Sharma et al., A comprehensive study of artificial neural networks. Int. J. Adv. Res. Comput. Sci. Softw. Eng. **2**(10), 278–284 (2012)

10. S. Gopal, *Artificial Neural Networks for Spatial Data Analysis* (Boston, 1988)
11. C.-T. Lin, C.-M. Yeh, S.-F. Liang, J.-F. Chung, N. Kumar, Support-vector-based fuzzy neural network for pattern classification. IEEE Trans. Fuzzy Syst. **14**, 31–41 (2006). https://doi.org/10.1109/TFUZZ.2005.861604
12. Y. Maali, A. Al-Jumaily, Self-advising support vector machine. Knowl.-Based Syst. **52**, 214–222 (2013). https://doi.org/10.1016/j.knosys.2013.08.009
13. T.K. Sajja, H.K. Kalluri, A deep learning method for prediction of cardiovascular disease using convolutional neural network. Rev. Intel. Artif. **34**(5), 601–606 (2020). https://doi.org/10.18280/ria.340510
14. J. Takalo-Mattila, J. Kiljander, J.P. Soininen, Inter-patient ECG classification using deep convolutional neural networks, in *2018 21st Euromicro Conference on Digital System Design, 8491848, IEEE Institute of Electrical and Electronic Engineers, Euromicro Conference on Digital System Design, DSD 2018, Prague, Czech Republic*, ed. by N. Konofaos, M. Novotny, A. Skavhaug, (IEEE, 2018), pp. 421–425. https://doi.org/10.1109/DSD.2018.00077
15. K.C. Sajja, A. Sweid, F. Al Saiegh, et al., Endovascular robotic: feasibility and proof of principle for diagnostic cerebral angiography and carotid artery stenting. J. Neuro Intervent. Surg. **12**, 345–349 (2020)
16. G. Taye, H.-J. Hwang, K. Lim, Application of a convolutional neural network for predicting the occurrence of ventricular tachyarrhythmia using heart rate variability features. Sci. Rep. **10**, 6769 (2020). https://doi.org/10.1038/s41598-020-63566-8
17. A. Azdani, K.D. Varathan, Y.K. Chiam, et al., A novel approach for heart disease prediction using strength scores with significant predictors. BMC Med. Inform. Decis. Mak. **21**, 194 (2021). https://doi.org/10.1186/s12911-021-01527-5
18. J. Heaton, Ian Goodfellow, Yoshua Bengio, and Aaron Courville: Deep learning: The MIT Press, 2016, 800 pp, ISBN: 0262035618. Genet. Program Evolvable Mach. **19**, 305–307 (2017). https://doi.org/10.1007/s10710-017-9314-z
19. Y. LeCun, L. Bottou, Y. Bengio, P. Haffner, Gradient-based learning applied to document recognition. Proc. IEEE **86**(11), 2278–2323 (1998). https://doi.org/10.1109/5.726791
20. A. Krizhevsky, I. Sutskever, G. Hinton, ImageNet classification with deep convolutional neural networks. Adv. Neural Inf. Proces. Syst. **25**, 84 (2012). https://doi.org/10.1145/3065386

# Chapter 3
# On the Performance of Codebook Designs on PD-SCMA Cognitive Radio Networks

**Simon Chege, Foued Derraz, Tom Walingo, and Yassin Bendimrad**

## 3.1 Introduction

The impact of non-orthogonal multiple access (NOMA) technologies, particularly on facilitating the 5G requirements of massive connectivity, ultra-high capacity and data rate, ultra-low latency, and robust interference management, has been immensely appreciated. NOMA multiplexes multiple users on a single resource element (RE) in power or code domains or their integration [1]. One possible integration is the hybrid NOMA technique namely power domain sparse code multiple access (PD-SCMA) scheme whose feasibility and performance are well investigated in [2–5]. At the code domain, PD-SCMA employs uniquely designed SCMA [6] codebooks to differentiate users using the same time–frequency resources. The multiplexing performance of such a scheme can be enhanced not only through deployment of sophisticated resource allocation schemes and low-complexity robust multi-user detectors but also prudent design of multi-user multi-dimensional codebooks [7]. On the other hand, cognitive radio networks (CRaNs), in the recent past, have been envisioned to provide enhanced and intelligent wireless spectrum

The work was supported in part by The ACADEMY Intra Africa Mobility Project, University of Tlemcen, Algeria and The Centre for Radio Access and Rural Technologies (CRART), University of KwaZulu Natal, South Africa.

S. Chege (✉) · T. Walingo
Discipline of Electrical, Electronic and Computer Engineering, University of KwaZulu Natal, Durban, South Africa

F. Derraz · Y. Bendimrad
Department of Telecommunications Engineering, University of Tlemcen, Tlemcen, Algeria

 
I. Woungang, S. K. Dhurandher (eds.), *5th International Conference on Wireless, Intelligent and Distributed Environment for Communication*, Lecture Notes on Data Engineering and Communications Technologies 174,
https://doi.org/10.1007/978-3-031-33242-5_3

efficiency. CRaNs are based on principles of spectrum sensing and dynamic spectrum access, where secondary users (SUs) sense the spectrum occupied by primary users (PUs), and intelligently adapt their operating parameters to access a spectrum band in an opportunistic or collaborative manner while keeping the interference temperature limit within the threshold [8]. The application of the hybrid NOMA in CRaN promises to enhance the spectral, connectivity, latency, and throughput requirements of 5G networks and is an active research.

The development of PD-SCMA CRaN system is a multi-stage design: CRaN resource allocation (RA), codebook design, and multi-user detector (MUD). The RA stage involves the design of sparse spreading signatures for the users, codebook assignment, user clustering, and power allocation for optimal PU–PU, PU–SU, and SU–SU interference mitigation. The second design step involves codebook design on top of the spreading signatures. In PD-SCMA, incoming user bits are mapped to multi-dimensional constellations (MDCs) in the code domain. Subsequently, adequate MDC design is instrumental in its performance. At the receiver, an optimal MUD with low complexity order is required for successful detection.

Primarily, codebooks generation is a two-fold process [9]: designing a multi-dimensional mother constellation (MC) and performing user-specific operations (USO) on MC to generate codebooks, where in PD-SCMA, USO is performed on both layer and RE dimensions. The original general rule on codebook design appeared in [10] where a MC was first generated by Cartesian product of two quadrature amplitude modulation (QAM) constellations and then USO using different permutation and complex conjugate operations [11] is performed. However, the permutation set used was not optimal leading to sub-optimality in power diversity gain among different codebooks.

Scanning through the open literature, several approaches to codebook designs are reported in [12, 13] for downlink and [14, 15] for uplink systems. Constellation rotation-based codebooks designed by constructing sub-constellations and employing Latin generator matrix for enhanced shaping gain are proposed in [12]. In [13], interleaving is proposed to maximize the minimum Euclidean distance (MED) among codewords of a codebook. Apart from phase rotation, coordinate interleaving, and permutation operations, a Trellis coded modulation (TCM)-based design is proposed in [14]. The design in [15] optimizes the mutual information and cut-off rate of an equivalent multiple-input multiple-output (MIMO) system. The bit error rate (BER) performance of the low-density signature (LDS) employing constellation rotation-based MDC designs outdoes others [9]. Contrastingly, few user/layer-specific operations have been designed for the uplink SCMA systems since they may lose its importance when UE transmits a codeword through a single layer. However, applying the multi-layer superposition transmission, the superposed codewords are transmitted through the independent channels, so the layer-specific operations need to be sophistically designed to mitigate inter-layer interferences.

The behavior of M-point MDCs on uplink SCMA is tied to the key performance indicators (KPIs) considered in the design process for different channel models [7]. In this chapter, we consider MDCs proposed in [16], [10], [17], [18], and [19]. The bit error rate (BER) performance of these MDCs is evaluated for low-rate PD-

SCMA systems over different channel scenarios through extensive simulations. To enhance the decoding experience and maintain uniqueness of users in both the power and code domains, a USO based on the diagonal $\Delta_k$ is employed [11]. Simulation results show significant performance enhancement for MDCs satisfying the KPIs. Moreover, the proposed USO enhances the receiver experience for the low-rate LTE turbo-coded system.

## 3.2 System Model

An uplink PD-SCMA-based underlay CRaN system is considered. The system model comprises a centralized macro-base station (MBS) serving a set of $\mathcal{U} = \{1, \cdots, U\}$ randomly distributed Pus and an underlaid set of $\mathcal{F} = \{1, \cdots, F\}$ low-power access points (APs), serving a set of $\mathcal{J} = \{1, \cdots, J\}$ uniformly distributed SUs. Similar to the conventional SCMA, a PD-SCMA transmitter operates $L$ layers (of set $\mathcal{L}$) and $N$ orthogonal resource elements (REs), where $N < L$. Each layer is assigned to only $d_v < N$ REs out of the $N$ REs. In other words, each layer spreads its data over $d_v$ REs. A layer is constructed by drawing select codewords from each user of SU set $\mathcal{J}$ and clustered PUs set $\mathcal{V}_{CB}$, ($|\mathcal{V}_{CB}| = V$, $\mathcal{V}_{CB} \in \mathcal{U}$). This implies that a layer constitutes $D = (V + 1)$ user symbols and $L = J$. Particularly, the PUs, SUs, and APs are equipped with a single antenna for simplicity in analysis. The total network bandwidth $B$ is equally shared by the $N$ REs. Figure 3.1 illustrates the PD-SCMA block. Under the constraint that no two layers should be assigned

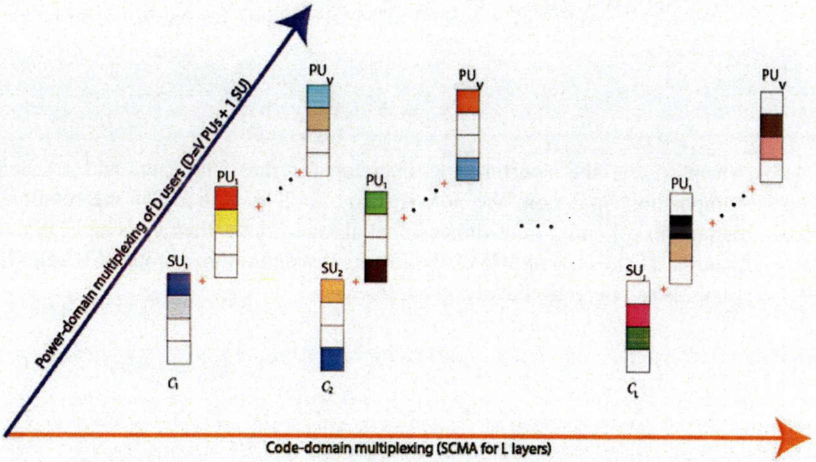


**Fig. 3.1** PD-SCMA block with $D = V + 1, V \in V_{CB}$ superimposed MUEs, $L$ layers/codebooks and $N = 4$ REs

all the same RUs for an affordable complexity order, a system is fully loaded if $\lambda = D \times \binom{L}{d_v}$.

A PD-SCMA transmitter combines V-BLAST encoding to obtain branch multiplexed signals, forward error correction (FEC) for correcting random error through introduction of redundancy and interleaving to resist consecutive errors through scattering the data stream. In the code domain, symbol mapping and the spreading are combined together for the SCMA encoder. The input bits are directly mapped to multi-dimensional complex domain codewords selected from a predefined codebook set. An SCMA encoder maps **b** incoming bits such that $\mathbb{B}^{log_2 M} \longmapsto \mathcal{S}, \mathbf{s} = g(\mathbf{b})$ with $\mathbf{s} \in \mathcal{S} \subset \mathbb{C}^N$ and $|\mathcal{S}| = M$. The $N$-dimensional complex codeword is sparse with $N - d_v$ non-zero entries. The encoder can be redefined to consider the mapping for only the $d_v$ constellation points of the codeword. Define a mapping from $\mathbb{B}^{log_2 M}$ to $\mathcal{C}$ given by

$$\mathbb{B}^{log_2 M} \longmapsto \mathcal{C}, \mathbf{c} = g(\mathbf{b}), \tag{3.1}$$

where **c** denotes the $d_v$-dimensional complex constellation point defined within the mother constellation set $\mathcal{C} \subset \mathbb{C}^{d_v}$. Then an SCMA encoder can be given as

$$f :\equiv \mathbf{W}g, i.e., \mathbf{s} = \mathbf{W}g(\mathbf{b}), \tag{3.2}$$

where the binary mapping matrix $\mathbf{W} \in \mathbb{B}^{L \times d_v}$ maps the $d_v$-dimensional complex codeword of mother constellation to a $N$-dimensional codeword of SCMA codebook. Observe that matrix **W** contains $N - d_v$ all-zero rows, and all the codewords of codebook contain zero in the same $N - d_v$ dimensions. We can get the matrix **W** by inserting $N - d_v$ row zero into $\mathbf{I}_{d_v}$.

From (3.1) and (3.2), a special SCMA codebook for one user can be expressed as

$$\mathbf{s}_j = \mathbf{W}_j(\Delta_j \cdot g)(\mathbf{b})_j, \tag{3.3}$$

where $\Delta_j$ is the constellation operator for the $j$-th user, and $(\cdot)$ denotes a composition operation. We note that $\Delta_j = \mathbf{E}_j\mathbf{1}$ with $\mathbf{E}_j$ as the rotation matrix defined in [6] and **1** is $N$-dimensional all one vector; then $\mathbf{s}_j$ is an LDS codeword that is a special case of SCMA. Following this definition, encoded SU and PU vector symbols can be, respectively, given by

$$\begin{aligned} \mathbf{s}^{SU_j} &= [s_1^{SU_j}, \cdots, s_K^{SU_j}]^T, \\ \mathbf{s}^{PU_v} &= [s_1^{PU_v}, \cdots, s_K^{PU_v}]^T \end{aligned} \tag{3.4}$$

with entries $s_k^{SU_j} = 0$ and $s_k^{PU_v} = 0$ at $N - d_v$ constellation points. Each codebook can be utilized by one user, like in conventional SCMA or several users superimposed in the codebook, as with PD-SCMA system by allocating users with distinct power levels.

After codebook assignment of the SUEs and power allocation, the MUE symbols in (3.4) are then clustered to get $\mathcal{V}_{CB}$ for pairing with a SUE assigned to a codebook using schemes in [5]. At each transmission period, a codeword vector is selected from each user codebook to constitute a layer $\mathbf{X}_l$ that can be expressed as

$$\mathbf{X}_l = \left[\mathbf{x}_l^{SU_j}\mathbf{x}_l^{PU_v}\cdots\mathbf{x}_l^{PU_V},\right] \in \mathbb{C}^{K\times M}. \tag{3.5}$$

The entries $\mathbf{x}_l^{SU_j} = \sqrt{P_l^{SU_j}}\cdot\mathbf{s}^{SU_j}$ and $\mathbf{x}_l^{PU_v} = \sqrt{P_l^{PU_v}}\cdot\mathbf{s}^{PU_v}$ and $P_l^{SU_j}$ and $P_l^{PU_v}$ are the normalized SU and PU power levels, respectively. The users in $\mathbf{X}_l$ are then multiplexed in power domain resulting to the layer message vector $\mathbf{x}_l \in \mathbb{C}^{K\times 1}$.

The received signal vector can be expressed as $\mathbf{y} = \left[y_1 y_2\cdots y_N\right]^T \in \mathbb{C}^{N\times 1}$ over the PD-SCMA block after the synchronous layer multiplexing, where $y_n$ as the received symbol on the $n^{th}$ RE is given as

$$\mathbf{y}_n = \sum_{l=1}^{L} diag(\mathbf{h}_l^n)\mathbf{x}_l + \mathbf{z}^n, \tag{3.6}$$

where $\mathbf{h}_l^n = \left[h_{l,SU_j}^n h_{l,PU_1}^n \cdots h_{l,PU_V}^n\right]$. Through the $n-th$ RE, $h_{l,SU_j}^n$ and $h_{l,PU_v}^n$ denote SU and PU channel coefficients averaged over the $d_v$ in each layer $l$, respectively.

In SCMA, the matrices $\mathbf{W}_j$ map each user to the resources in a sparse allocation. Given that the matrices are fixed and independent of the selected codewords, the layer RE assignment is also fixed during the transmissions. As a result, the SCMA structure can then precisely be described by a graph structure as illustrated in Fig. 3.2 and its respective indicator matrix defined by $\mathbf{F} = (f_1, \cdots, f_L)$. The

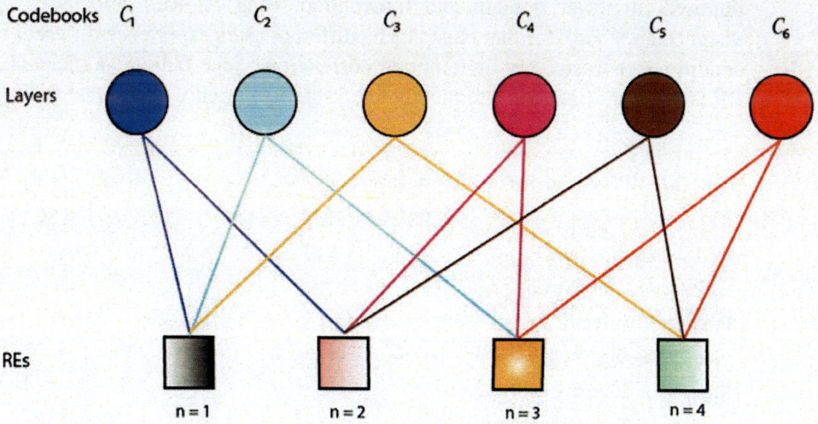


**Fig. 3.2** PD-SCMA factor graph representation with $L = 6$ and $N = 4$ REs

layer node $l$ and resource node $n$ are connected if and only if $(\mathbf{F})_{n,l} = 1$. Using matrix $\mathbf{F}$, we can easily describe the graph structure of SCMA. Define the sets $\vartheta_l = \{n : f_{n,l} = 1\}$, of the resources where layer $l$ spreads its codeword, and $\varphi_n = \{l : f_{n,l} = 1\}$, of the users colliding in resource $n$. The relationships $d_v = |\vartheta_l|$ and $d_f = |\varphi_n|$ are respectively named the spreading degree of layer $l$ and collision degree of resource $n$. The indicator matrix of Fig. 3.2 can be given as

$$\mathbf{F}_{4\times 6} = \begin{bmatrix} 1\,1\,1\,0\,0\,0 \\ 1\,0\,0\,1\,1\,0 \\ 0\,1\,0\,1\,0\,1 \\ 0\,0\,1\,0\,1\,1 \end{bmatrix}. \tag{3.7}$$

The PD-SCMA system users spread their data symbols over $d_v$ REs. The REs can experience several channel mode cases, namely: Uncoded fading where each user observes the same channel coefficients over its REs (FSC), uncoded fading where each user observes independent channel coefficients over its REs (FIC), coded fast fading where each user observes the same channel coefficients over its REs (FFSC), coded fast fading where each user observes independent channel coefficients over its REs (FFIC), coded quasi-static fading where each user observes the same channel coefficients over its REs (SFSC), and lastly, coded quasi-static fading where each user observes independent channel coefficients over its REs (SFIC) [7].

The design process of $M$-point $d_v$-dimensional complex constellations $\mathcal{C}^+$ considers key performance indicators (KPIs) applicable under the channel scenarios of transmission. The authors in [7] highlight these KPIs, namely: Euclidean distance, Euclidean Kissing Number, Product Distance, Product Kissing Number, Modulation Diversity Order, the Number of Distinct Points, and Bit Labeling. Table 3.1 represents the consideration KPI in each channel mode.

In the uplink PD-SCMA transmission, users (MUEs and SUEs) are superimposed in power domain and transmitted using $d_v$ REs associated with the assigned codebook for that cluster. The different users experience different channel conditions characterized by temporal correlation of the fading coefficients at each RE at different times. Consequently, the users' experience coded fast fading where

**Table 3.1** KPIs considered in different channel modes

| KPI | FSC | FIC | FFSC | FFIC | SFSC | SFIC |
|---|---|---|---|---|---|---|
| Euclidean distance, $d^2_{E,min}$ | ✓ | | ✓ | | ✓ | |
| Euclidean Kissing Number, $\tau_E$ | ✓ | | ✓ | | ✓ | |
| Product Distance, $d^2_{P,min}$ | | ✓ | | ✓ | | ✓ |
| Product Kissing Number, $\tau_P$ | | ✓ | | ✓ | | ✓ |
| Modulation Diversity Order, $L$ | | ✓ | | ✓ | | ✓ |
| Number of Distinct Points, $N_d$ | ✓ | ✓ | ✓ | ✓ | ✓ | ✓ |
| Bit Labeling | ✓ | ✓ | ✓ | ✓ | ✓ | ✓ |

each user observes independent channel coefficients over its REs in which most of the KPIs should be put into consideration for efficient MDC design.

## 3.3 Multi-dimensional Constellation Designs

In this section, we highlight state of the art $M$-point $d_v$-dimensional constellations proposed for SCMA systems and applicable for PD-SCMA systems. We consider a system with $M \in \{4, 16\}$ and $d_v = 2$ REs.

### 3.3.1 LDS Codebooks

The LDS codebooks are designed by spreading a symbol $x$, taken from a quadrature-amplitude modulation (QAM) set of size $M$ over some sparse signature $\mathbf{f}$. This can be observed as repeating the $M$-QAM constellation point over the 2 REs [16]. The complex projections in each dimension of this LDS codebook are illustrated in Figure 3.3.

### 3.3.2 TM-QAM Codebooks

Proposed in [10] and later patented in [17], the TM-QAM MDC design for multiple layers involves two steps. First, a $d_v$-dimensional complex MDC with a good Euclidean distance profile is designed. Second, a unitary rotation matrix is applied to the base constellation to maximize the minimum product distance.

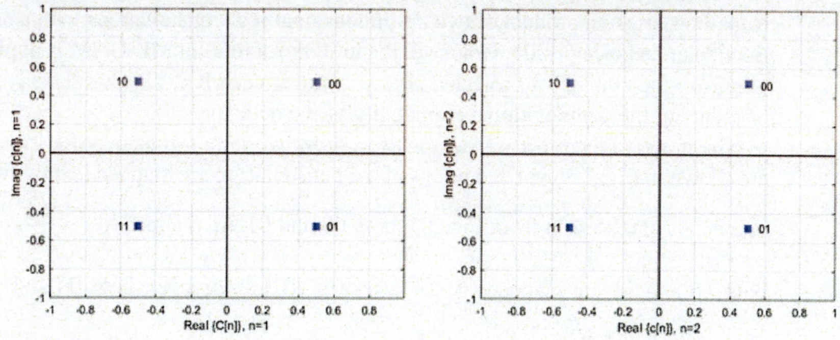


(a) (b)

**Fig. 3.3** LDS codebook with $(M, N) = (4, 2)$. **(a)** LDS, $n = 1$. **(b)** LDS, $n = 2$

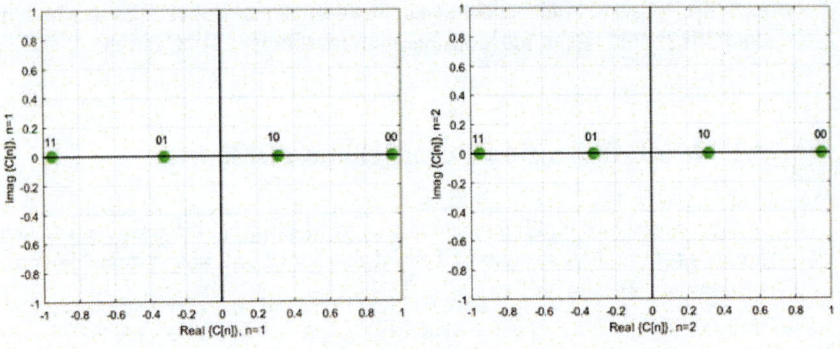


(a) (b)

**Fig. 3.4** TM-QAM codebook with $(M, N) = (4, 2)$. **(a)** T4-QAM, $n = 1$. **(b)** T4-QAM, $n = 2$

The complex projections onto two dimensions for $M = 4$ are shown in Fig. 3.4. Each user maps the incoming bitstreams onto the constellation with green circles in Fig. 3.4 in order to send them over RE1 and RE2. With the rotated constellation, various sets of operators such as phase rotations [17], Latin square rotation criterion [20], structured approach design of spread sequences [11], and diversity-order-based rotation technique [21] can be applied to build multiple TM-QAM-based sparse codebooks for multiple layers of SCMA.

### 3.3.3 ML-QAM Codebooks

The $M$-point constellation design proposed in [18] is based on the [10] and [17] but with a low number of projections. Similar to TM-QAM, the design process involves a shuffling method to establish a $d_v$-dimensional complex constellation from the Cartesian product of two $d_v$-dimensional real constellations with a desired Euclidean distance profile. However, the unitary rotation in ML-QAM is applied to minimize the number of projected points, $N_d$ over each RE. This has the advantage of reducing the computational complexity of the detector.

The projections of ML-QAM for $M = 4$ on two dimensions over RE1 and RE2 are shown in Fig. 3.5, where 4 points of the constellation are mapped to only $N_d = 2$ points. As can be seen from Fig. 3.5, both 00 and 01 are mapped to $(\frac{-\sqrt{2}}{2}, 0)$, and both 10 and 11 are mapped to $(\frac{\sqrt{2}}{2}, 0)$ over RE1. Moreover, both 00 and 10 are mapped to $(0, \frac{-\sqrt{2}}{2})$, and both 01 and 11 are mapped to $(0, \frac{\sqrt{2}}{2})$ over RE2. The lowered number of projections results in reduced computational complexity from $4^{d_f}$ to $2^{d_f}$.

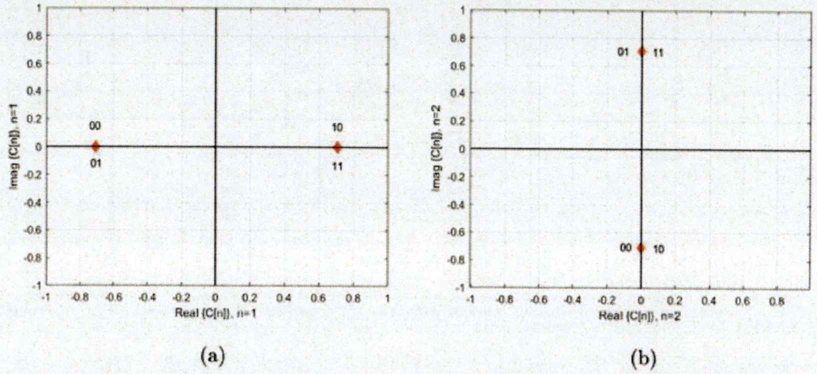


**Fig. 3.5** ML-QAM codebook with $(M, N) = (4, 2)$. **(a)** 4L-QAM, $n = 1$. **(b)** 4L-QAM, $n = 2$

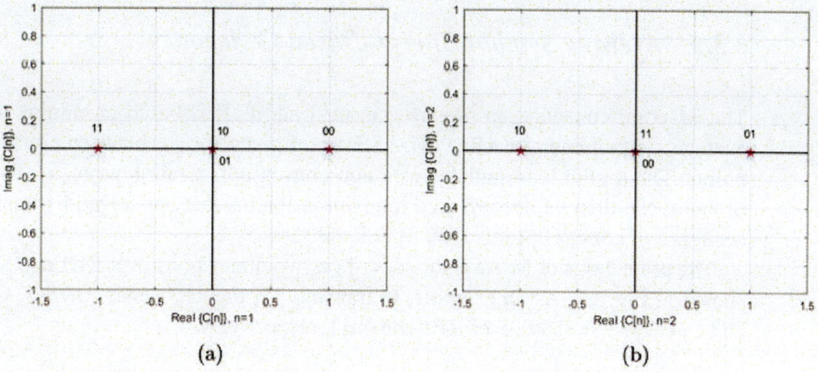


**Fig. 3.6** MC-QAM codebook with $(M, N) = (4, 2)$. **(a)** 4C-QAM, $n = 1$. **(b)** 4C-QAM, $n = 2$

### *3.3.4 MC-QAM Codebooks*

The $M$-point circular QAM constellation (MC-QAM), proposed in [18], is based on the analysis of the signal space diversity (SSD) for multiple-input multiple-output (MIMO) systems over Rayleigh fading channels in [13, 14]. Similar to ML-QAM, MC-QAM targets to obtain a low number of projections per complex dimension.

The projections of ML-QAM for $M = 4$ on two dimensions over RE1 and RE2 are shown in Fig. 3.6. The 4 points of the constellation are mapped to only $N_d = 3$ points, therefore minimizing the complexity from $4^{d_f}$ to $3^{d_f}$. With Gray labeling, both 01 and 10 are mapped to $(0, 0)$, 00 mapped to $(1, 0)$ and 11 mapped to $(-1, 0)$ for transmitting over RE1. On the other hand, to transmit over RE2, both 00 and 11 are mapped to $(0, 0)$, 01 mapped to $(1, 0)$, and 10 mapped to $(-1, 0)$.

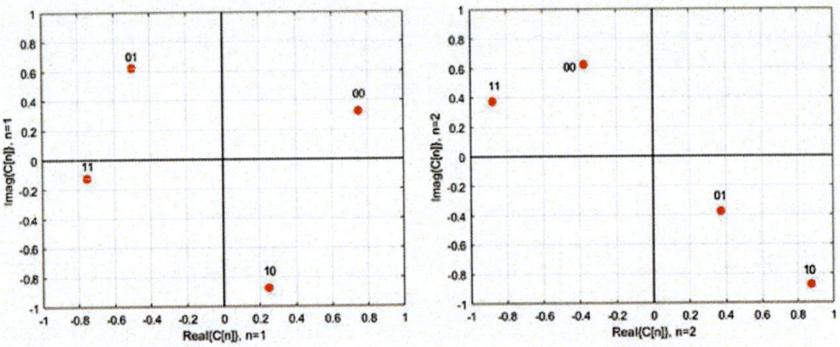

(a) (b)

**Fig. 3.7** M-ASE codebook with $(M, N) = (4, 2)$. (**a**) M-ASE, $n = 1$. (**b**) M-ASE, $n = 2$

### 3.3.5 *Average Symbol Energy-Based Codebooks*

The $M$-point constellation proposed in [19] and [22] is based on minimizing the average symbol energy (ASE) (M-ASE) for a given $d_{E,min}$ between constellation points. The design is formulated as a non-convex optimization problem. The main problem can then be decomposed into sub-problems that are solved by applying a sequence of convex optimization techniques.

The projections of M-ASE for $M = 4$ on two dimensions over RE1 and RE2 are shown in Fig. 3.7. As an example, to transmit 11, the user sends $(0.7543, 0.3852)$ and $(-0.3993, 0.3509)$ over RE1 and RE2, respectively.

## 3.4 Performance Evaluation of the MDCs on PD-SCMA

In this section, through extensive simulations, the performance of those constellations applied on PD-SCMA scheme over different channel models is evaluated. Since we consider the model in the uplink, each user observes independent channel coefficients over its REs. Consequently, we only show results for FIC, FFIC, and SFIC channel models. The KPIs corresponding to the considered MDCs are provided in [7] and given in Table 3.2. As observed in Table 3.1, most of the KPIs have significant impact on the BER performance of uncoded systems. Moreover, the behavior of the MUD at different SNRs affects the performance.

The BER performance of uncoded PD-SCMA system in FIC is illustrated in Fig. 3.8. At low SNRs, 4-LDS and 4c-QAM outperform other MDCs. The M-ASE achieves improved performance at high SNRs than the other considered MDCs at the same KPIs. M-ASE exhibits the highest $d^2_{E,min}$ compared with other 4-point constellations since it is designed for AWGN channels. This notwithstanding, 4-ASEs also have the highest $\tau_E$ that affects its performance especially at low SNRs.

**Table 3.2** KPIs considered in different channel modes

| | $d^2_{E,min}$ | $\tau_E$ | $d^2_{P,min}$ | $\tau_P$ | $L$ | $N_d$ | Gray-labeled |
|---|---|---|---|---|---|---|---|
| T4-QAM | 2 | 2 | 0.64 | 2 | 2 | 4 | Yes |
| 4L-QAM | 2 | 2 | 2 | 2 | 1 | 2 | Yes |
| 4C-QAM | 2 | 2 | 1 | 2 | 1 | 3 | Yes |
| 4-ASE | 2.67 | 3 | 0.29 | 0.5 | 2 | 4 | No |
| LDS | 2 | 2 | 1 | 2 | 2 | 4 | Yes |

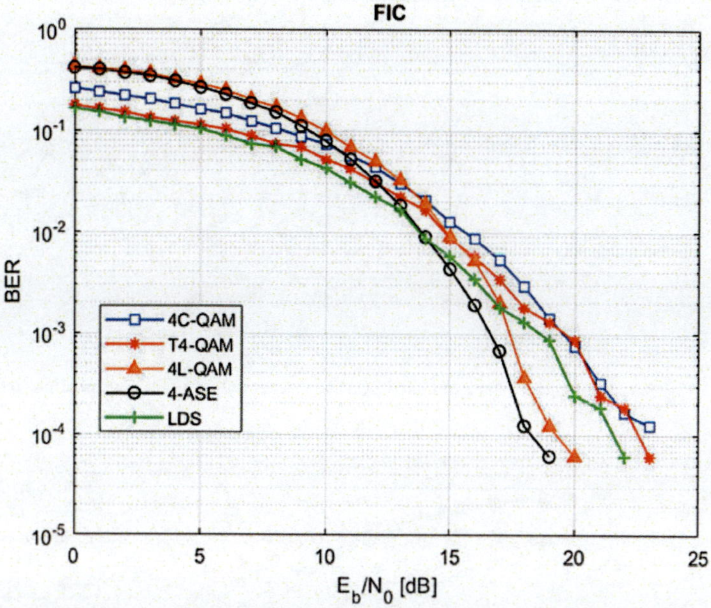


**Fig. 3.8** BER performance of uncoded SCMA systems with 4-point constellations over FIC

Since in FIC, each user manifests independent channel coefficients over the $d_v = 2$ REs, it is expected that the channel diversity order will be $>1$. This translates to a change of $>10$ dB in SNR per a change of one decade in BER.

In Fig. 3.9, the BER performance of coded PD-SCMA system in FFIC channel scenario is illustrated. As addressed in 3.1, the type of bit labeling, the SNR operating region, $N_d$, $L$, $d_{p,min}$, and $tau_p$ are the KPIs for the FFIC scenario. The 4-LDS and 4C-QAM outperform the others in the FFIC channel scenario. Among these constellations, 4-LDS and 4C-QAM have the high $d^2_{P,min}$ and the same $\tau_P$. As such, both 4-LDS and 4C-QAM outperform the others. Moreover, compared with the other constellations with $\tau_P$, the performance of 4L-QAM is deteriorated by its lowest $L$.

Lastly, we compare the BER performance of coded PD-SCMA system in SFIC channel scenario as illustrated in Fig. 3.10. Note that since in quasi-static fading

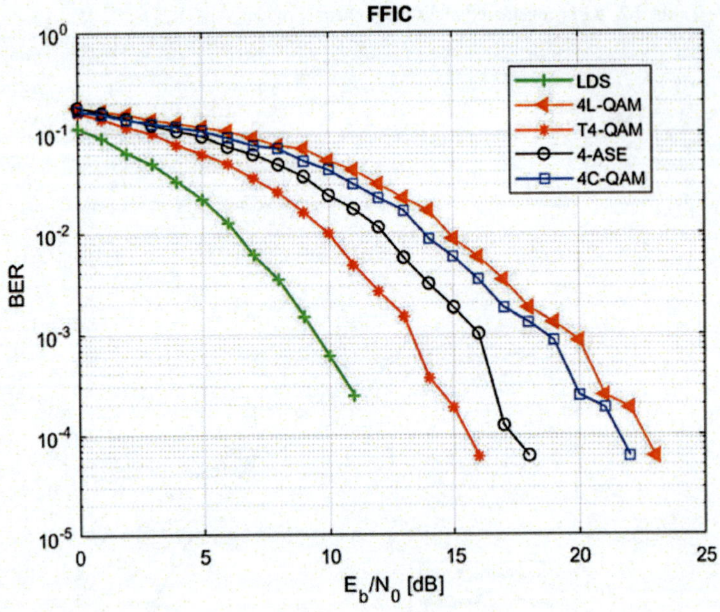


**Fig. 3.9** BER performance of coded SCMA systems with 4-point constellations over FFIC

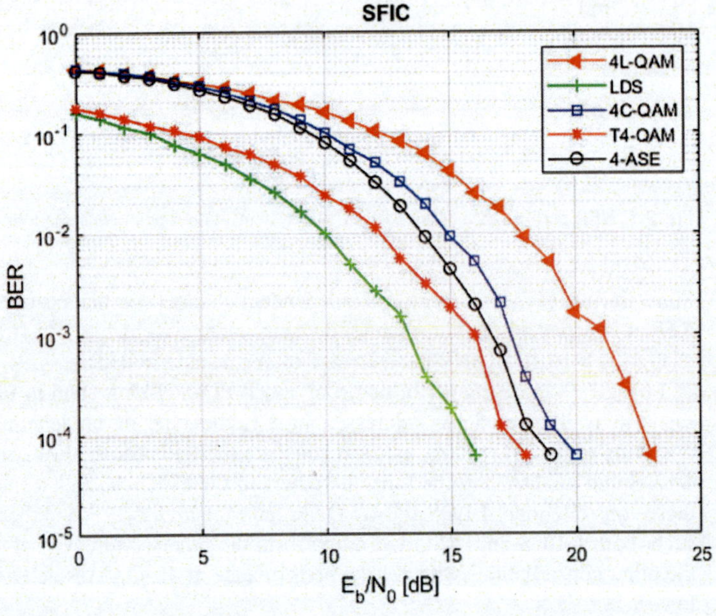


**Fig. 3.10** BER performance of coded SCMA systems with 4-point constellations over SFIC

scenarios each channel coefficient is constant for the duration of transmission of the whole codeword, the system performs poorly. Hence, in quasi-static fading scenarios, higher SNR regions are mainly of interest. The 4-LDS and T4-QAM exhibit better performance than the other constellations at low and high SNRs. They are closely followed by 4-ASE, while 4L-QAM trails behind due to its bit labeling. Further, it can be observed that the trends of some MDCs for a target BER of $10^{-2}$ are consistent with the trends in BER performance of uncoded systems at their corresponding SNRs $\approx$17.2 dB.

## 3.5 Conclusion

In this chapter, the behavior of MDCs on the performance of PD-SCMA systems for CRaN networks is analyzed for low-rate systems. It is observed that different constellations perform differently over various channel scenarios. Consequently, in order to optimize the performance in a certain channel scenario, a proper MDC can be designed using the specific KPIs for that scenario. In a PD-SCMA scenario, where multiple users traverse the code and power domains simultaneously, an optimal and dynamic design of MDCs and the associated user-specific rotation design to be applied in both domains are under consideration as part of our future work. Besides, the performance of the MDCs for high-rate coded systems is a possible research direction.

## References


1. Y. Cai, Z. Qin, F. Cui, G.Y. Li, J.A. McCann, Modulation and multiple access for 5G networks. IEEE Commun. Surv. Tutor. **20**(1), 629–646 (2018)
2. M. Moltafet, N. Mokari, M.R. Javan, H. Saeedi, H. Pishro-Nik, A new multiple access technique for 5G: power domain sparse code multiple access (PSMA). IEEE Access **6**, 747–759 (2017)
3. T. Sefako, T. Walingo, Biological resource allocation algorithms for heterogeneous uplink PD-SCMA NOMA networks. IEEE Access **8**, 194 950–194 963 (2020)
4. S. Chege, T. Walingo, Energy efficient resource allocation for uplink hybrid power domain sparse code nonorthogonal multiple access heterogeneous networks with statistical channel estimation. Trans. Emerg. Telecommun. Technol. **32**(1), e4185 (2021)
5. S.K. Chege, T. Walingo, Multiplexing capacity of hybrid PD-SCMA heterogeneous networks. IEEE Trans. Veh. Technol. 1–1 (2022). http://doi.org/10.1109/TVT.2022.3162304
6. H. Nikopour, H. Baligh, Sparse code multiple access, in *2013 IEEE 24th Annual International Symposium on Personal, Indoor, and Mobile Radio Communications (PIMRC)* (IEEE, 2013), pp. 332–336
7. M. Vameghestahbanati, I.D. Marsland, R.H. Gohary, H. Yanikomeroglu, Multidimensional constellations for uplink SCMA systems–a comparative study. IEEE Commun. Surv. Tutor. **21**(3), 2169–2194 (2019)
8. S.M. Almalfouh, G.L. Stüber, Interference-aware radio resource allocation in OFDMA-based cognitive radio networks. IEEE Trans. Veh. Technol. **60**(4), 1699–1713 (2011)

9. B.F. da Silva, D. Silva, B.F. Uchôa-Filho, D. Le Ruyet, A multistage method for SCMA codebook design based on MDS codes. IEEE Wirel. Commun. Lett. **8**(6), 1524–1527 (2019)
10. M. Taherzadeh, H. Nikopour, A. Bayesteh, H. Baligh, SCMA codebook design, in *2014 IEEE 80th Vehicular Technology Conference (VTC2014-Fall)* (IEEE, 2014), pp. 1–5
11. J. Van De Beek, B.M. Popovic, Multiple access with low-density signatures, in *GLOBECOM 2009–2009 IEEE Global Telecommunications Conference* (IEEE, 2009), pp. 1–6
12. Y. Zhou, Q. Yu, W. Meng, C. Li, SCMA codebook design based on constellation rotation, in *2017 IEEE International Conference on Communications (ICC)* (IEEE, 2017), pp. 1–6
13. D. Cai, P. Fan, X. Lei, Y. Liu, D. Chen, Multi-dimensional SCMA codebook design based on constellation rotation and interleaving, in *2016 IEEE 83rd Vehicular Technology Conference (VTC Spring)* (IEEE, 2016), pp. 1–5
14. J. Bao, Z. Ma, Z. Ding, G.K. Karagiannidis, Z. Zhu, On the design of multiuser codebooks for uplink SCMA systems. IEEE Commun. Lett. **20**(10), 1920–1923 (2016)
15. S.-C. Lim, H. Park, Codebook optimization for the superposition transmission of uplink SCMA systems. IEEE Trans. Veh. Technol. **67**(10), 10 112–10 117 (2018)
16. R. Hoshyar, F.P. Wathan, R. Tafazolli, Novel low-density signature for synchronous CDMA systems over AWGN channel. IEEE Trans. Signal Process. **56**(4), 1616–1626 (2008)
17. M.T. Boroujeni, H. Nikopour, A. Bayesteh, M. Baligh, System and method for designing and using multidimensional constellations, US Patent 9,509,379, 29 Nov 2016.
18. T. Metkarunchit, SCMA codebook design base on circular-QAM, in *2017 Integrated Communications, Navigation and Surveillance Conference (ICNS)* (IEEE, 2017), pp. 3E1–1
19. M. Beko, R. Dinis, Designing good multi-dimensional constellations. IEEE Wirel. Commun. Lett. **1**(3), 221–224 (2012)
20. J. Dénes, A.D. Keedwell, in *Latin Squares: New Developments in the Theory and Applications*, vol. 46 (Elsevier, Amsterdam, 1991)
21. J. Boutros, E. Viterbo, Signal space diversity: a power-and bandwidth-efficient diversity technique for the Rayleigh fading channel. IEEE Trans. Inf. Theory **44**(4), 1453–1467 (1998)
22. H. Nikopour, M. Baligh, Systems and methods for sparse code multiple access, US Patent 9,240,853, 19 Jan 2016

# Chapter 4
# T-AES- and ECC-Based Secure Data Communication in Peer-to-Peer Networks

**Mukesh Kumar, Kuldeep Singh Jadon, and Nitin Gupta**

## 4.1 Introduction

In today's time, data are everything and data security is a biggest concern. Due to the lack of a centralized system that has to be protected, P2P networking poses an intriguing security challenge [1]. Rather, the nodes need to cooperate with one another to guarantee proper and safe functioning. Unfortunately, the presence of malignant nodes in the network is a natural consequence of deploying software on a public network like the Internet. These nodes will attempt to interfere with the system in some way or undermine it so that they can benefit from it. P2P systems have to be built to function appropriately even under conditions like these. Thus data security is also very important, and data need to be transferred securely through a secure channel [2, 3].

Most of the earlier work for data transfer have mainly focused on two approaches: first, by using encryption technique and other using cloud security solutions. For encryption, symmetric or asymmetric technique can be used [4]. Asymmetric encryption provides good level of security, but it is very slow and can only be used to encrypt lesser data. On the other hand, symmetric encryption can be used to encrypt large data; however, it is also not able to achieve reasonable efficiency. MAES [5] is also implemented to reduce the time taken to encrypt the data.

MAES [5] provides a new key implementation process, the right-shift key technique, and the elimination of the Mix-Column process was introduced. The main emphasis was provided on the key modification technique. Keys were morphed

---

M. Kumar (✉) · K. S. Jadon · N. Gupta
Department of Computer Science and Engineering, National Institute of Technology, Himachal Pradesh, India
e-mail: nitin@nith.ac.in



I. Woungang, S. K. Dhurandher (eds.), *5th International Conference on Wireless, Intelligent and Distributed Environment for Communication*, Lecture Notes on Data Engineering and Communications Technologies 174,
https://doi.org/10.1007/978-3-031-33242-5_4

and changed, but cipher text was unchanged in round operation. Thus, MAES is more efficient and with less complexity along with a minimum execution time that surpasses the conventional AES propositions. But MAES can also be improved further by applying the threading, while encrypting the data that has been done in the proposed work and simulation result shows that T-AES is much more efficient than MAES.

Third-party provider or cloud service provider gives a high level of security. However, data are very sensitive information, and there is risk of third party taking the benefit of our data or use it for their own purpose. Moreover, if they are encrypting the data, then the key is shared between us and third party and they can decrypt the data whenever required. Therefore, a solution for faster data encryption and transmission by achieving the high level of security is highly desired.

In the proposed work, T-AES is proposed to encrypt the data, and ECC [6] has been used for the AES key encryption. T-AES substitutes two bytes parallelly, which reduces the time taken by the substitution bytes (S-box) and increases the efficiency of the algorithm. AES is symmetric encryption technique [7], which means that it uses same key for encryption and decryption. Data encryption is done with the help of AES key, and this key should be shared with the receiver as well for data encryption. Therefore, securing the AES key is also necessary. Diffie–Hellman key exchange protocol (DHKEP) [8] is used to generate the shared key between sender and receiver, and AES key is encrypted using ECC.

The main contributions of the proposed work are as follows:

1. T-AES with ECC is implemented for data transfer from one machine to another, by achieving high efficiency and without enhancing the security level.
2. Threading is applied on AES that increased the overall efficiency of the algorithm by 42% and reduces the total transmission time to less than a second.
3. DHKEP is used to generate the shared key, which will be used to encrypt and decrypt the AES key on sender and receiver sides, respectively.
4. Higher throughput is achieved, and the overall execution time is reduced by the factor of 1.72.

The remaining parts of this work are structured in the following manner. Section 4.2 gives an overview of the strategies utilized for secure data transfer as well as related research. In Sect. 4.3, data encryption and decryption are discussed. Performance evaluation of the proposed work has been done in Sect. 4.4 followed by the conclusions.

## 4.2 Literature Review

Numerous studies have been conducted on the modification of AES cryptography and its manipulation on cloud senders using big data. A. Sachdeva et al. [9] discussed unpredictable secrecy as a security method with AES-based storage requirements. AES was chosen because it requires less storage and executes more quickly than other approaches. Fadul et al. [10] proposed two secret keys as a safe technique, with the second key (extra) being utilized for both encryption and decryption. According to the test, this improved security while keeping the performance measure near to the original AES. To protect the cloud computing paradigm, similar research effort with a focus on cloud computing is presented in Akhil et al. [11]. Furthermore, the field of cloud computing is investigated.

Arab et al. [12] first built the encryption key using Arnold chaos [13] system, and the encryption is performed by combining the chaos system and AES that is called modified AES. This technique is applied to encrypt the images. The substitution and column integration processes in this algorithm were changed to be replaced by other operations, which resulted in the suggested method being quicker than the first AES one and even immune to all different types of threats.

Musa et al. [14] had concluded that the major route of transmission is P2P, which creates a risk because of the nature of the exchange. The effect that P2P networks have on data security will determine whether or not they become commonplace in traditional computer settings. If enterprises use peer-to-peer networks in the same way that scanners and mails are used in today's world, then the danger of attacks from insiders will be increased. The usage of P2P network connection exposes the system to the possibility of a virus being introduced into it.

Wagh et al. [15] specified rigorous security measures to protect a data owner's files, while they are stored in the cloud infrastructure. They were primarily concerned with preserving the data's confidentiality and authenticity. The employment of public key, hash, and secret key cyphers between the sender and recipient guaranteed a safe cloud environment. Conducting an online aptitude exam for qualified pupils and offering security for the same are two future expansions.

Makhija et al. [16] offered an introduction to cloud computing, which is anticipated to be embraced by governments, manufacturers, and academicians in the very near future. The author also presented a broad overview of all available cloud data security measures, as well as ways for assuring authentication methods through third-party administrators (TPA). Arman et al. [5] have combined the AES with ECC, and the input text file is converted into encrypted form using AES encryption, but the key is created using ECC [17]. To obtain the original text file, the receiver used that key to decode the text file that was submitted to the sender in encrypted form. Finally, numerous criteria such as storage need, encryption time, decryption time, avalanche impact, and correlation were used to analyze AES encryption with ECC [18].

As discussed above, various algorithms are proposed to secure the data. The danger increases while transmitting the data through P2P networks. In this proposed work, T-AES-based data encryption scheme is proposed, which results in improved efficiency and higher throughput compared to previous work. Security of AES key is also kept in mind, and ECC is used to secure the AES key. The shared key is generated using Diffie–Hellmann protocol that is used by ECC to encrypt the AES key. By using threads on AES, it reduced the overall time of the encryption process and data can be shared instantly among multiple peers, achieving low latency.

## 4.3 Proposed Work

In this section, the data transfer from sender to receiver using T-AES and ECC is described. First, AES is implemented using threads to make the encryption process faster, and AES key is secured with the help of asymmetric encryption, i.e., ECC and DHKEP are used to get the shared key. The various steps are described below.

### 4.3.1 *Diffie–Hellman Key Exchange Protocol*

Figure 4.1 describes the Diffie–Hellman key exchange [19]. Receiver and sender both have their public and private keys. Sender's public key and receiver's private key are combined together and shared key is generated. Now, this shared key is common between sender and receiver.

Let us say public keys available to receiver and sender are P,G and private key selected by sender is $a$ and by receiver is $b$.

Next, the key generated at sender side is x=$G^a mod P$, and receiver side is y=$G^b mod P$. After generating the key, receiver and sender exchange the key and secret key is generated.

```
Key received at sender = y
Key received at receiver = x
```

$$\texttt{sender secret key, } K_s = y^a mod P$$

$$\texttt{receiver secret key, } k_r = x^b modp$$

$$\texttt{Algebraically } k_s == k_r$$

Hence, shared key is generated. This is used by the sender and receiver to encrypt and decrypt the AES key, respectively.

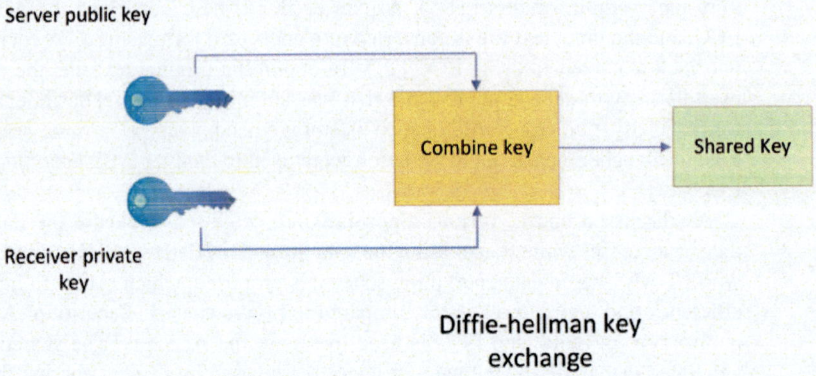


**Fig. 4.1** Diffie–Hellman key exchange

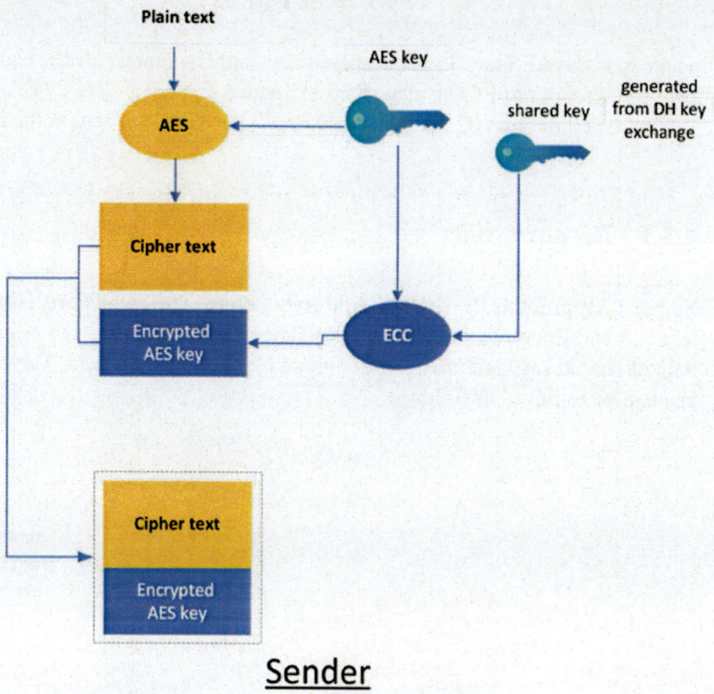


**Fig. 4.2** Sender side encryption process

### *4.3.2 Sender Side*

Figure 4.2 shows data transfer from sender to receiver in encrypted form with the help of AES and ECC. The sender is having the data in plain-text format that are encrypted using AES and sent to the receiver. The receiver decrypts the data and performs required operations on it. Next, AES key is encrypted using ECC with the help of shared key that is generated using Diffie–Hellman key exchange protocol between sender and receiver.

Let us denote the plain text with $P$, $K$ denotes the AES key, and $C$ denotes the cipher text, and then the mathematical equation for encryption can be represented as

$$C = E(P, K), \tag{4.1}$$

where $E(P, K)$ is the AES encryption of $P$ using key $K$. The working of AES encryption E(P,K) is described in the encryption step of section D. Next, AES key is encrypted using ECC and is given as follows:

$$EA = ECC(K, k_s), \tag{4.2}$$

where K is the AES key, $k_s$ is the shared key, and EA represents the encrypted AES key. The working of ECC is described in Section E.

Finally, cipher text (C) and encrypted AES key (EA) are sent to the receiver.

### *4.3.3 Receiver Side*

Figure 4.3 represents the receiver side architecture. The receiver receives the cipher text $(C)$ and Encrypted AES key $(EA)$. The original AES key is retrieved with the help of shared key, and finally, the original plain text is retrieved. The original AES key can be retrieved as follows:

$$K = ECC(EA, k_r), \tag{4.3}$$

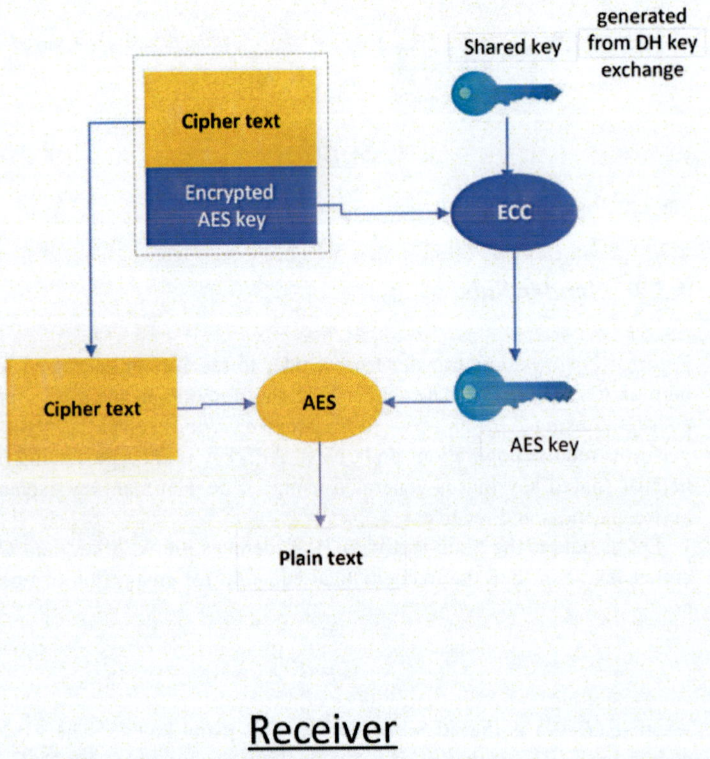


**Fig. 4.3** Receiver side data decryption

where $EA$ is the encrypted AES key that is transferred from the sender and $k_r$ is the shared key. After getting the original AES key (K), cipher text is decrypted to the plain text as follows:

$$P = D(C, K), \tag{4.4}$$

where $P$ is the original plain text, and $D(C, K)$ represents the decryption function that is applied on the cipher text $(C)$ and AES key $(K)$. The decryption step is described in detail in the decryption process of Section D.

### 4.3.4 *The Proposed T-AES Scheme*

The processes for encrypting and decrypting data using T-AES are outlined in this section:

1. **Encryption:** For encryption process, the function is represented by

$$F_c = P \oplus K, \tag{4.5}$$

   where $F_c$ is a cipher text function, $P$ is the plain text, and $K$ is the key. Encryption method follows the number of steps that are described below:

   a) A plain text and 128-bit user defined key are taken for the encryption process where keys and plain text are divided into a $4 \times 4$ matrix.
   b) Then the plain text is XOR with the initial round key given as

$$C_0 = P \oplus K_0, \tag{4.6}$$

   where $C_0$ is the cipher text, P is the plain text, and $K_0$ represents the initial key.
   c) After performing the XOR, all data are rearranged into a $4 \times 4$ matrix. For key and cipher text, total 16 bytes of data are there.
   d) For the second round, initial key is altered. Left shift is performed on the first row of $4 \times 4$ matrix, and right shift on the second row is done two times. The S-box is used to change the third row. Additional technique is used to change the fourth row, and our second key is generated by following this process, $K_1$. These steps are done using multi-threading to make this process faster.
   e) Now, cipher text and second key $K_1$ are XOR-ed together for getting the $C_1$.

$$C_1 = C_0 \oplus K_1. \tag{4.7}$$

   f) Similar process (altering technique and XOR operations with the each newly generated cipher text) will be followed to generate the 9 more keys. Next, after the 10 rounds, we will have keys and cipher text, i.e., $K = (0, 1 \ldots 9)$ and C = (0,1…8).

g) To get the $10^{th}$ key $K_{10}$, same altering process is followed except the $4^{th}$ row. Fourth row is kept unchanged and that is how we will achieve our last key ($K_{10}$).
h) The main cipher text will be generated by XOR-ing the last key with the last cipher text. This last key and cipher text will be shared with the receiver for the decryption.

$$C_{10} = C_9 \oplus K_{10}. \tag{4.8}$$

Figure 4.4 depicts the encryption procedure in detail.

2. **Decryption:** Function representation of the decryption process is as follows:

$$F_p = C \oplus K, \tag{4.9}$$

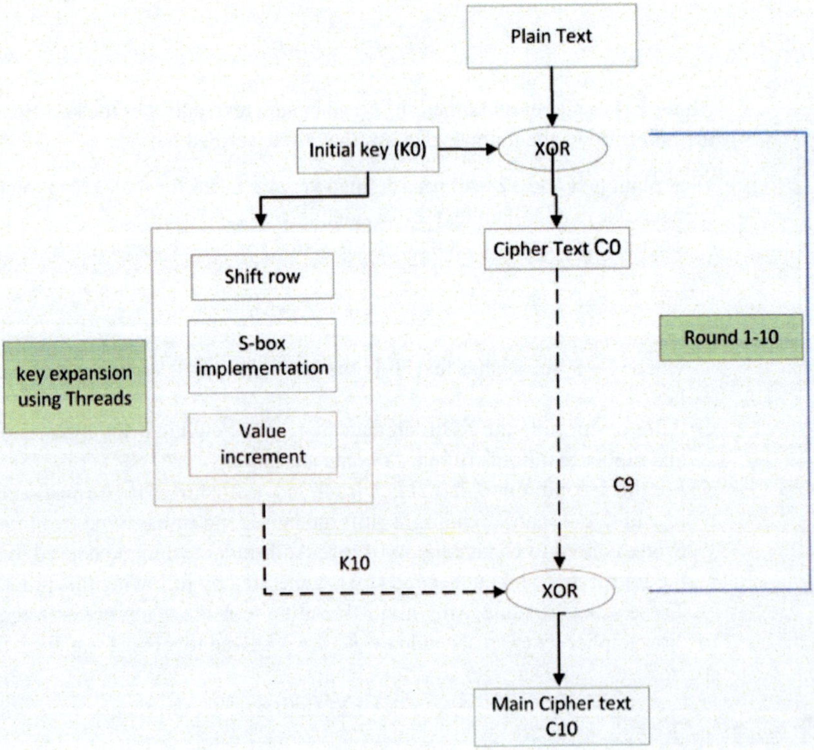


**Fig. 4.4** T-AES encryption process flowchart

where $F_p$ is a plain-text function, C represents the cipher text, and K denotes key. Decryption process steps are described below:

a) We have been provided with the cipher text ($C_{10}$) and the key ($K_{10}$) from the sender. This cipher text and key will be converted into a 4 × 4 matrix for the decryption.
b) Now, we will XOR them, and we would have $C_9$, i.e., tenth cipher text.

$$C_9 = C_1 0 \oplus K_1 0. \tag{4.10}$$

c) Now, $K_{10}$ will go through the reverse process, i.e., vice versa of encryption process would be followed, i.e., right shift the first row, left shift the second row twice, we will substitute third row with the value of the inverse S-box that we have implemented. That is how we will get the $K_9$.
d) $K_9$ will be XOR-ed with the cipher text $C_9$ that produces the cipher text ($C_8$).

$$C_8 = C_9 \oplus K_9. \tag{4.11}$$

e) Now, we have to achieve the $K_8$. So we will again tweak the $K_9$. For the upper three rows, same process will be followed as mentioned in the above step (step 3), and decrement logic will be added up to the fourth row to make it properly.
f) Now, $K_8$ will be XOR-ed with the cipher text ($C_8$) to achieve the next cipher text ($C_7$). The keys will also go through the same process as described in the earlier steps. Until we receive the last cipher text and key, it will cipher text will be XOR-ed with the key, and new key will be generated. Hence, at last, we will have K = (10,9,. . . 0) and C = (10,9. . .).
g) Now, we will XOR the last cipher text ($C_0$ and key ($K_0$) to get the our main plain text. Decryption process is the exact opposite of the encryption process. Therefore, we will have our plain text (P) at the end.

$$P = C_0 \oplus K_0. \tag{4.12}$$

Figure 4.5 describes the steps followed by decryption process.

### 4.3.5 ECC

In ECC, a fundamental flaw in RSA is the sudden growth in key size, which is compensated by smaller keys. ECC [20] is also more efficient for a multitude of reasons. To begin with, smaller keys indicate that during an SSL handshake [21], less data must be communicated from the sender to the recipient. Furthermore, EC requires less processing power (CPU) and memory, resulting in much faster response times and throughput on Internet senders once implemented. Perfect

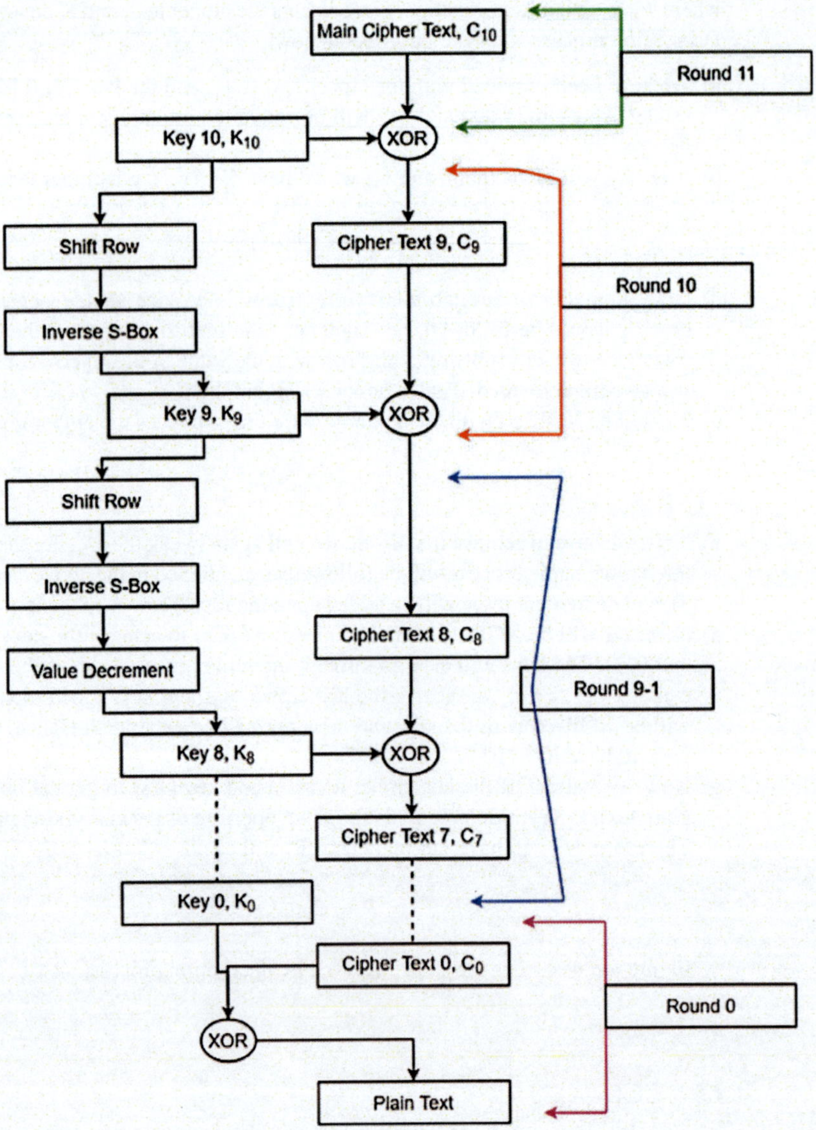


**Fig. 4.5** T-AES decryption process flowchart

forward secrecy is an important feature of employing ECC. A 160-bit ECC key is about the same size as a 1024-bit RSA key.

The ECC algorithm works in the following way. Let $E_q(a, b)$ be an elliptic curve with parameters a,b and q be a prime number or an integer of the form $2^m$. Let F be the point on the elliptic curve whose order is large value of n. Then

**Sender side key generation:**

$n_a$ is the selected private key, $n_a < n$

Public key $P_a = n_a * G$

**Receiver side key generation:**

$n_b$ is the selected private key, $n_b < n$

Public key $P_b = n_b * G$

Sender secret key, $K = n_a * P_b$

Receiver secret key, $K = n_b * P_a$

a) **ECC encryption:** Let M be the message. First, convert the message M into a point on elliptic curve, i.e., $P_m$. Choose a random positive integer k for encryption. For encryption, the public key of receiver is used, i.e., $P_b$. The cipher point will be $C_m = \{kG, P_m + kP_b\}$. Now this pair will be shared with the receiver.
b) **Decryption at receiver side:** For decryption, the private key of receiver is used. Multiply the first point in pair with the receiver secret key, i.e., $kG * n_b$. Now, subtract it from the second point in the pair.

$$= P_m + kP_b - kG * n_b,$$

and we know that $P_b = n_b * G$

$$= P_m + kP_b - K * P_b = P_m.$$

Hence, the receiver gets the original point.

The whole P2P secure data transfer using the proposed approach is summarized in Fig. 4.6.

## 4.4 Simulation and Results

In this section, results and comparison of AES performance is presented. The considered threaded AES for the encryption shows remarkable improvement and increases the efficiency of the algorithm.

In the simulation of the proposed work, the system configuration has RAM of 4 GB, processor i7, a hard disk of 64 GB, and a processor speed of 2.4 GHz.

Table 4.1 shows the different simulation parameters chosen for the simulation of the proposed work.

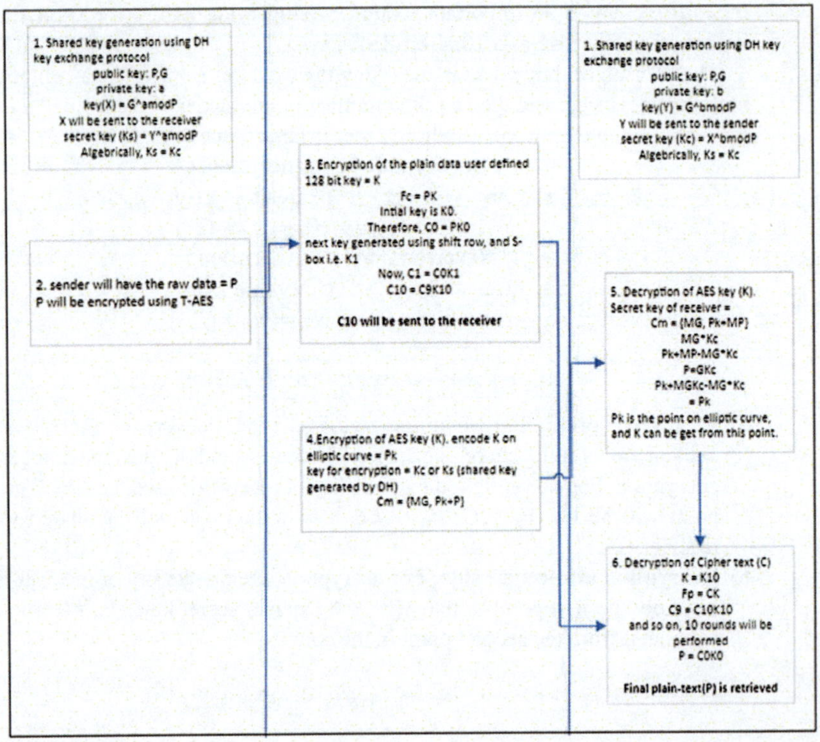


**Fig. 4.6** Flow of P2P secure data transfer

**Table 4.1** Simulation parameters

| S.No. | Parameters | Value |
|---|---|---|
| 1 | Key size | 128, 192, 256 |
| 2 | No. of dataset | 20 |
| 3 | CPU cores | 2,4,8 |
| 4 | No. of nodes | 4 |

Table 4.2 shows the total time taken during data transfer from sender to receiver. The transmission time (data packets transferred) is assumed as 0.1 s since it depends on the network speed, and generally, the data delivered over the network are very fast. For encryption and decryption time, we get the results from our algorithm. The table shows that after applying threading on AES, it took less than a second for the whole data transmission. Hence, the execution time is reduced by a factor of 1.72 (1.71/0.99) or T-AES is 42% more efficient than the AES [5] and MAES [5].

Figure 4.7 shows the transmission time of different algorithms in graphical representation. T-AES has shown the improvement over AES and MAES, and the total transmission time in T-AES is less than compared to other algorithms used in this proposed work.

**Table 4.2** Data transfer using AES-128

| | Time (in seconds) | | | |
|---|---|---|---|---|
| Algorithms | Encryption time | Transmission time | Decryption time | Total transmission time |
| AES | 0.88 | 0.1 | 0.88 | 1.71 |
| MAES | 0.65 | 0.1 | 0.65 | 1.31 |
| T-AES | 0.49 | 0.1 | 0.49 | 0.99 |

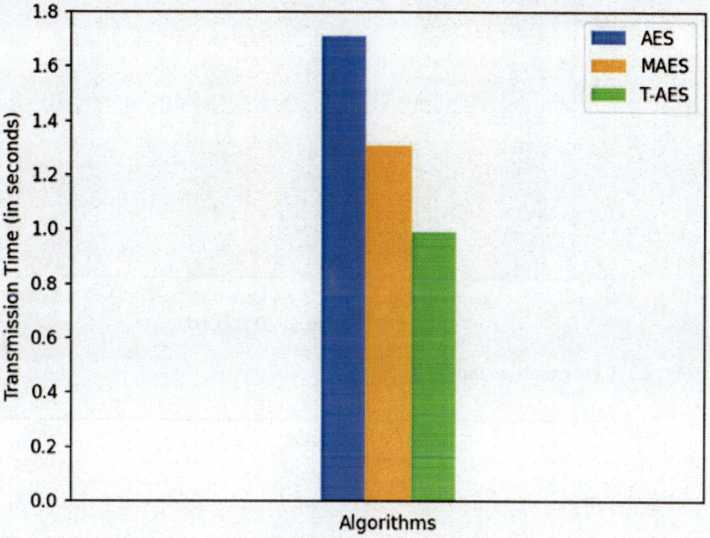


**Fig. 4.7** Total Transmission Time (128-bit key)

Next, the execution time is calculated and is performed in python on 128-bit block of data. We have prepared 20 different random datasets that are used on different algorithms to evaluate their performances. Figure 4.8 shows the normalized execution time based on the CPU for 128 bit. The proposed algorithm is tested on multi-core CPUs, and the result shows that it is 42% more efficient than the existing MAES approach on 8-core CPU.

Next, the throughput is calculated. To calculate the throughput, we average the execution time of each algorithm. The results are shown in Fig. 4.9. As it can be observed, the proposed T-AES scheme has achieved high efficiency as compared to AES and MAES.

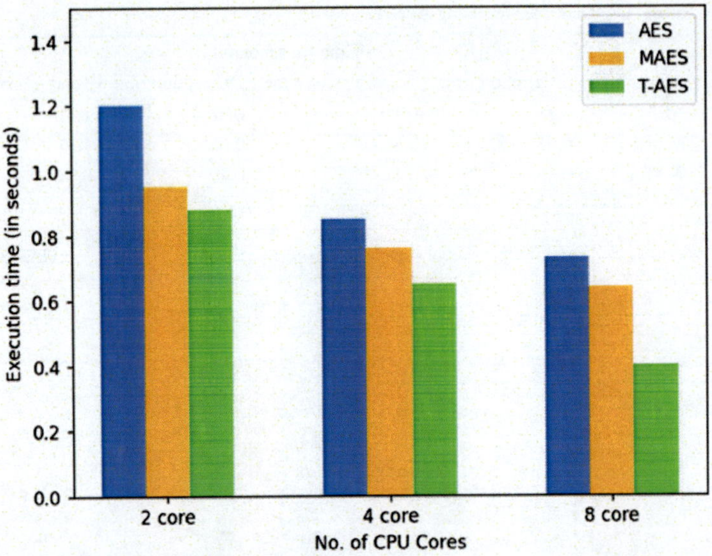


**Fig. 4.8** CPU execution time

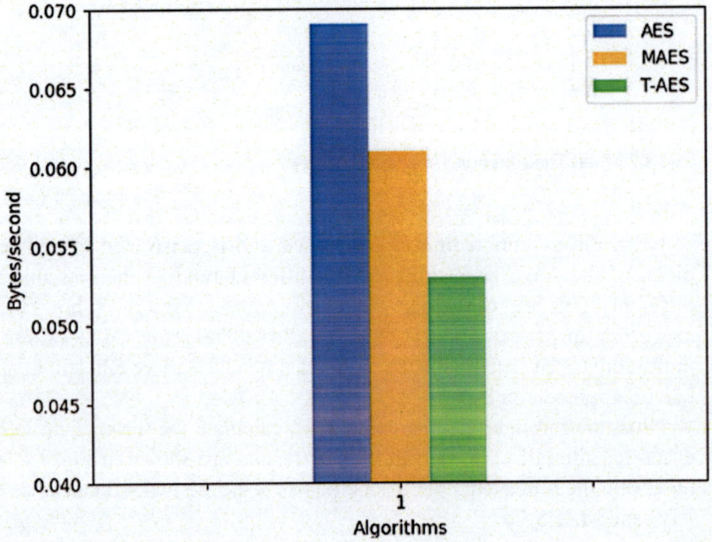


**Fig. 4.9** Throughput

### *4.4.1 Security Analysis*

The proposed modified AES using threads is based on AES-128. A N-bit size cipher has a key space of $2^N$ bytes. Therefore, there are total $2^{128} = 3.4 * 10^{38}$ alternative keys. It will take minimum $5.4 * 10^{18}$ years to get the exact key. It can be observed that the threaded AES is still very difficult to break by the attackers. Further, from the above evaluation, it can be concluded that the proposed approach achieved high level of security as well as fast data transmission. Moreover, by adding threads in AES, the encryption process will be faster, and security of AES is also not compromised.

## 4.5 Conclusions

Data security is one of the main requirements at the present time, and AES cryptographic systems are highly strong; however, the encryption/decryption process is also quite difficult. Encrypting data in larger quantities take longer time due to its complexity. As a result, a quicker encryption technique with uncompromised security was necessary. To address the aforementioned flaws, the AES technique is updated in this work. The right-shift approach and the elimination of Mix-Column made the criteria simple and effective at the same time. The primary modification approach received the most attention. In round operations, keys were mutated and altered, while the cipher text remained intact. All these modifications are done using threads, which made the algorithm more faster. The objective was to keep costs and computations to a minimum. Moreover, the data clearly supported the hypothesis. The efficiency gain was significant due to the security trade-off. As shown in the results, the proposed solution is more efficient and sophisticated, with a minimum execution time that outperforms traditional AES proposals.

In the future work, digital signature standard (DSS) or TLS/SSL protocol can be used to secure the AES key instead of ECC. The proposed algorithm has used threading, which can be replaced by multiprocessing or multiprogramming in the future work to make the algorithm more efficient, and then the algorithm can be used to transfer the satellite or healthcare data with much more improved speed and high security.

## References

1. N.R. Pradhan, A.P. Singh, N. Kumar, M. Hassan, D. Roy, A flexible permission ascription (FPA) based blockchain framework for peer-to-peer energy trading with performance evaluation. IEEE Trans. Industr. Inform. **18**, 2465–2475 (2021)
2. A. Erceg, Information security: threat from employees. Teh. Glas. **13**(2), 123–128 (2019)

3. S.S. Srivastava, N. Gupta, R. Jaiswal, Modified version of playfair cipher by using 8×8 matrix and random number generation, in *IEEE 3rd International Conference on Computer Modeling and Simulation* (2011)
4. R.B. Marqas, S.M. Almufti, R.R. Ihsan, Comparing symmetric and asymmetric cryptography in message encryption and decryption by using AES and RSA algorithms. J. Xian Univ. Archit. Technol. **12**, 3110–3116 (2020)
5. S. Arman, T. Rehnuma, M. Rahman, Design and implementation of a modified AES cryptography with fast key generation technique, in *2020 IEEE International Women in Engineering (WIE) Conference on Electrical and Computer Engineering (WIECON-ECE)* (IEEE, 2020), pp. 191–195
6. A. Shantha, J. Renita, E.N. Edna, Analysis and implementation of ECC algorithm in lightweight device, in *2019 International Conference on Communication and Signal Processing (ICCSP)* (IEEE, 2019), pp. 0305–0309
7. S. Rajesh, V. Paul, V.G. Menon, M.R. Khosravi, A secure and efficient lightweight symmetric encryption scheme for transfer of text files between embedded IoT devices. Symmetry **11**(2), 293 (2019)
8. K. Amine, Diffie-Hellman key exchange through steganographied images. Brasilia **10**(1), 147–160 (2018)
9. A. Sachdev, M. Bhansali, Enhancing cloud computing security using AES algorithm. Int. J. Comput. Appl. **67**(9), 19–23 (2013)
10. I.M.A. Fadul, T.M.H. Ahmed, Enhanced security of Rijndael algorithm using two secret keys. Int. J. Secur. Appl. **7**(4), 127–134 (2013)
11. K. Akhil, M.P. Kumar, B. Pushpa, Enhanced cloud data security using AES algorithm, in *2017 International Conference on Intelligent Computing and Control (I2C2)* (IEEE, 2017), pp. 1–5
12. A. Arab, M.J. Rostami, B. Ghavami, An image encryption method based on chaos system and AES algorithm. J Supercomput. **75**(10), 6663–6682 (2019)
13. M.F.A. Elzaher, M. Shalaby, S.H. El Ramly, An Arnold cat map-based chaotic approach for securing voice communication, in *Proceedings of the 10th International Conference on Informatics and Systems* (2016), pp. 329–331
14. A. Musa, A. Abubakar, U.A. Gimba, R.A. Rasheed, An investigation into peer-to-peer network security using Wireshark, in *2019 15th International Conference on Electronics, Computer and Computation (ICECCO)*, (IEEE, 2019), pp. 1–6
15. K. Wagh, R. Jathar, S. Bangar, A. Bhakthadas, Securing data transfer in cloud environment. Int. J. Eng. Res. Appl. **4**, 91–93 (2014)
16. B. Makhija, V. Gupta, I. Rajput, Enhanced data security in cloud computing with third party auditor. Int. J. Adv. Res. Comput. Sci. Softw. Eng. **3**(2), 341–345 (2013)
17. T. Tomar, V. Kumar, Y. Kumar, et al. Implementation of elliptic–curve cryptography. Int. J. Electr. Eng. Technol. **11**, 178–189 (2020)
18. S. Sharma, V. Chopra, Analysis of AES encryption with ECC, in *Proceedings of the 17th International Interdisciplinary Conference on Engineering Science & Management* (Dubai, 2016), pp. 1–2
19. L.C. Huang, T.Y. Chang, M.S. Hwang, A conference key scheme based on the Diffie-Hellman key exchange. Int. J. Netw. Secur. **20**(6), 1221–1226 (2018)
20. D.B. Roy, D. Mukhopadhyay, High-speed implementation of ECC scalar multiplication in GF (p) for generic Montgomery curves. IEEE Trans. Very Large Scale Integr. (VLSI) Syst. **27**(7), 1587–1600 (2019)
21. T.M. Zaw, M. Thant, S. Bezzateev, User authentication in SSL handshake protocol with zero-knowledge proof, in *2018 Wave Electronics and its Application in Information and Telecommunication Systems (WECONF)* (IEEE, 2018), pp. 1–8

# Chapter 5
# Effective Fatigue Driving Detection by Machine Learning

**Hwang-Cheng Wang and Jia-Jun Zhuang**

## 5.1 Introduction

Fatigue driving is one of the main reasons for severe accidents. Statistics compiled by the Ministry of Transportation and Communications (MOTC) of Taiwan show that road traffic accidents in Taiwan have been growing year by year [1]. According to a research report by the Transport Research Institute of the MOTC in 2014, the costs of road traffic accidents in Taiwan are enormous, including medical costs, lost productivity, and degraded quality of life associated with the casualty. Based on the report, it is estimated that the average cost per person killed in a car accident is about $15.72 million, and the average cost per injured person is $1.19 million, which translates into an economic loss of NT$551.592.41 million attributed to traffic accidents in 2018 alone. This represents about 3.1% of the Gross Domestic Product (GDP) for that year [2], showing that traffic accidents have a profound impact on the economy. According to the MOTC, distracted and fatigue driving accounts for 20% of traffic accidents [3], which is a significant proportion. Data show that tired driving accounts for 16–20% of road crashes in Victoria, Australia [4]. To solve the problem of fatigue driving, it is necessary to determine whether a driver is fatigued or not accurately.

In this paper, we investigate the use of machine learning to determine whether a driver is fatigued using the condition of eyes and mouth as a clue. We investigate several approaches that have been used for the purpose and compare their merits and weaknesses.

H.-C. Wang (✉) · J.-J. Zhuang
National Ilan University, Yilan, Taiwan
e-mail: hcwang@niu.edu.tw



I. Woungang, S. K. Dhurandher (eds.), *5th International Conference on Wireless, Intelligent and Distributed Environment for Communication*, Lecture Notes on Data Engineering and Communications Technologies 174,
https://doi.org/10.1007/978-3-031-33242-5_5

According to [5–7], there are three main ways to obtain driver fatigue characteristics, which are based on vehicle parameters, based on human behavior, and based on physiological parameters. The acquisition of vehicle parameters needs to be based on the hardware integration of the vehicle itself, and the complexity of operation may also decrease the willingness to use the approach. Acquired parameters include the following:

- Real-time lane detection system
- Nine-axis sensing of a vehicle
- Time series analysis of steering wheel angular velocity

Human behavior mainly refers to the behavioral information obtained through noninvasive measurements, such as through in-vehicle cameras, environmental microphones, and other devices. The amount of data obtained from human behavioral information is relatively small, and machine learning is required to obtain fatigue-related features. Information in this category includes the following:

- Blink frequency increase
- Too much time with eyes closed
- Frequent yawning
- Continuous driving time
- Facial expression recognition
- Eyeball tracking [8]
- Head posture

The third type is based on physiological parameters collected by invasive sensors, including the following:

- Driver fatigue detection based on electroencephalography (EEG) [9]
- Heart rate variability wavelet analysis and SVM classifier [10]
- Pulse Sensors [11]
- Wearable driver drowsiness detection devices
- Wireless wearable devices
- Hybrid method using physiological characteristics [12]

Compared with the above two types, this type has the most fatigue information in the data collected, but it also requires the integration of invasive sensors, which may cause discomfort to users and provide a low incentive to use it.

The following are some of the methods used in the study: face triaxial rotation data, facial recognition techniques, mouth and eye closure, and information entropy of face motion. Our objective is to integrate these methods for inferring the results of driver fatigue through neural network models. The models are based on a common CNN model, one with several fully connected layers and the other incorporating long short-term memory (LSTM). We compared the proposed designs and showed that they both have excellent performance.

This paper is organized as follows. Section 5.1 provides the motivation and overview of the research. Related works are briefly reviewed in Sect. 5.2. The proposed method is described in Sect. 5.3. In Sect. 5.4, we first elaborate on ways to reduce the amount of data used in the training of machine learning models. Two machine learning structures are then described. Fatigue detection results using the two machine learning models are presented and discussed. Finally, Sect. 5.5 concludes this paper.

## 5.2 Related Works

### *5.2.1 Facial Feature Landmark Model*

According to [13, 14], Dlib is one of the common lightweight 68-point face marking methods. The principle behind it can be divided into two parts. One is to obtain face features by Histogram of Orientation Gradient (HOG), and the other is to do linear category differentiation by Support Vector Machine (SVM). Dlib has the advantage of minimal training data, but the re-accuracy rate, stability, and maximum detectable angle are poor.

MTCNN plays a major role in face recognition using deep learning [15]. Compared with the traditional Dlib algorithm, MTCNN can adapt to more complex environments, larger detection angles, accuracy, stability, etc. Its special algorithm optimization makes MTCNN computation speed comparable to Dlib algorithm. BlazeFace [16] is a lightweight real-time facial recognition model developed by Google for mobile devices, and its output contains 468 facial 3D annotation points (Fig. 5.1). Compared to the above two models, this model uses additional object tracking algorithms to perform facial recognition, making it more stable and improve the computational efficiency. One drawback of the method is that it is only applicable to a single photo containing little facial information. In comparison with the other methods, BlazeFace is more suitable for our research.

### *5.2.2 Facial Recognition Model: Facenet-Inception-Resnet*

FaceNet [17] is a framework introduced by Google, which first performs face detection on the input image, followed by face alignment, and then obtains facial features through the Resnet algorithm Feature Extraction. architecture [18]. Resnet was introduced by Microsoft in 2015, and the residual architecture proposed at that time solved the degradation problem caused by multilayer network training, enabling the learning of more complex features. Then, the core algorithm of FaceNet, Triplet Loss, is used to minimize the distance of an anchor and a positive of the same identity and maximize the distance between the anchor and a negative of a different identity. If the Euclidean distance between the feature vectors of two pictures is small, persons in two pictures are judged to be the same.

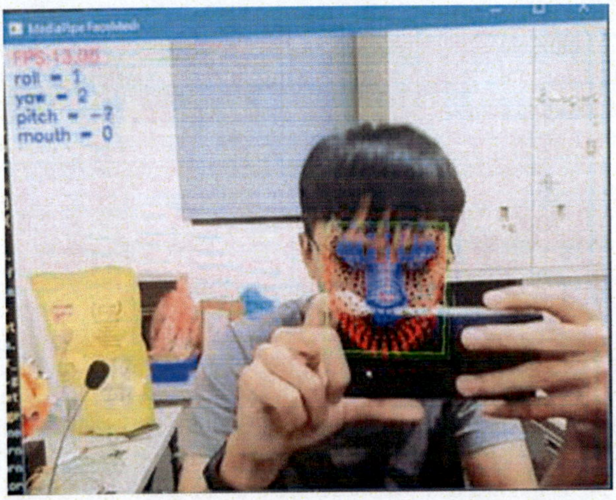


**Fig. 5.1** BlazeFace is able to recognize a face even when it is partially obscured

### 5.2.3 *Facial Motion Information Entropy Algorithm*

In most of the studies on fatigue feature acquisition, the basis for assessing the driver's state is through the analysis of a single frame or several frames, which are unsuitable for inferring long-time driving behaviors. Facial motion information entropy [19] can obtain a large amount of the driver's facial motion information for a long period of time and analyze the fatigue state based on a certain length of a sliding window.

### 5.2.4 *One-Dimensional Convolutional Neural Network (1D CNN)*

2D CNN is very effective in extracting important features in a large amount of data, such as image recognition and voice recognition, whereas 1D CNN can also be used to extract data features in one-dimensional space [20]. 1D CNN is very useful for extracting essential features from a large amount of driver's state sequence data, and the appropriate addition of convolutional layers can significantly prevent the network model from overfitting.

### 5.2.5 *Time Series Analysis Model: LSTM*

It is usually challenging to determine whether a driver is in a fatigued state by a single or small amount of picture data, and it is usually necessary to analyze a large amount of data with a temporal relationship to better determine the fatigue state [21]. The LSTM (long short-term memory) model is used to analyze and train the data. The LSTM is based on the RNN (Recurrent Neural Network), which is also used for temporal data processing [22]. LSTM is mainly used to solve the vanishing gradient and exploding gradient problems during the training process, which provides better performance on long time-series data training.

## 5.3 Proposed Method

### 5.3.1 *Thermal Imaging and MTCNN for Real-Time Face Recognition*

In this approach, we first capture the approximate human head outline through thermal imaging, which is then mapped to a photo taken by a camera to obtain a low-resolution image. Next, the five feature points of the human face are captured through MTCNN for real-time face recognition. This helps reduce the overall computational workload. The process is illustrated in Fig. 5.2.

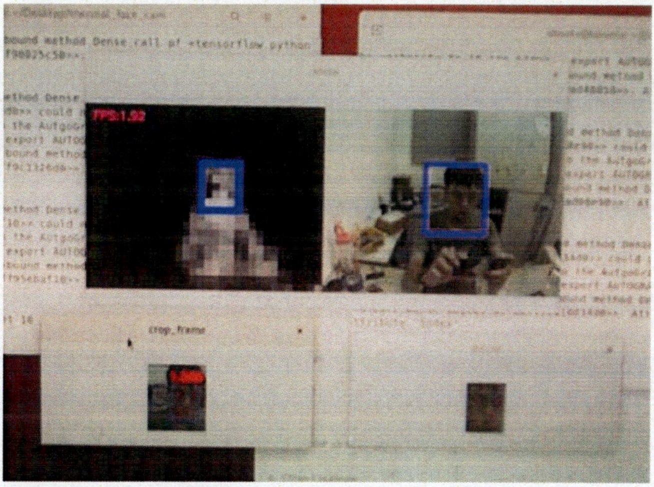

**Fig. 5.2** Thermal imaging and MTCNN for face recognition

**Table 5.1** CPU utilization of different methods

| Method | MTCNN face recognition | Thermal image acquisition | Hybrid |
|---|---|---|---|
| CPU usage | 45% | 4% | 16% |

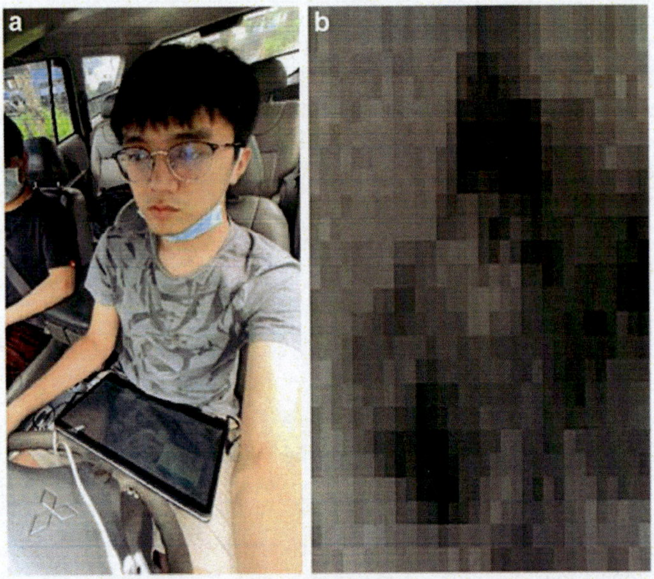


**Fig. 5.3** (**a**) Full color picture and (**b**) thermal imaging

In addition, we found that the FPS (frames per second) was low in practice, mainly because the I2C of the Raspberry Pi was limited in speed due to the system compatibility issue. To solve the problem, we use an additional MCU to forward the signal. In this setup, the Raspberry Pi is connected to the MCU via UART, and then the MCU is connected to the infrared camera (MLX90640) via I2C. The implementation allows us to achieve a higher FPS.

For a comparison of the performance of different methods, the test conditions are fixed at 8 FPS with a resolution of 320 × 240 pixels. Table 5.1 summarizes the results. The results show that using thermal imaging capture and MTCNN face recognition can significantly reduce the load of CPU.

When the test was conducted outdoors, the recognition accuracy was found to drop sharply. Since the outdoor temperature was around 30° on that day, the infrared camera captured images in which the human body temperature and the ambient temperature were blended (Fig. 5.3). The results suggest that the method has to be used with caution, taking into account the surrounding environment.

### 5.3.2 Obtaining Yawning and Eye-Opening/Closing Features

In reference to [21, 23], we first obtained 468 3D facial landmarks through the facial recognition model BlazeFace. We then computed EAR (eye aspect ratio) and MAR (mouth aspect ratio) values by Eq. (5.1).

$$\text{EAR or MAR} = \frac{\|p_1 - p_6\| + \|p_3 - p_5\|}{2\,\|p_1 - p_4\|}, \tag{5.1}$$

where ||•|| denotes the Euclidean distance between two points.

The landmarks used in the computing of EAR are illustrated in Fig. 5.4. Finally, machine learning is used to determine whether the target is in a fatigued state. The test results are shown in Fig. 5.5. Here, we use a simple LSTM model and the YAWDD (Yawning Detection Dataset) for yawning detection.

The YawDD dataset comprises a total of 322 driving videos in which the drivers exhibit yawning, talking, and mouth-closing behaviors. We use an LSTM model with parameters shown in Fig. 5.6. The input to our model is in the form of ($N$, 300, 1), where $N$ represents the amount of training data, 300 is the number of time units

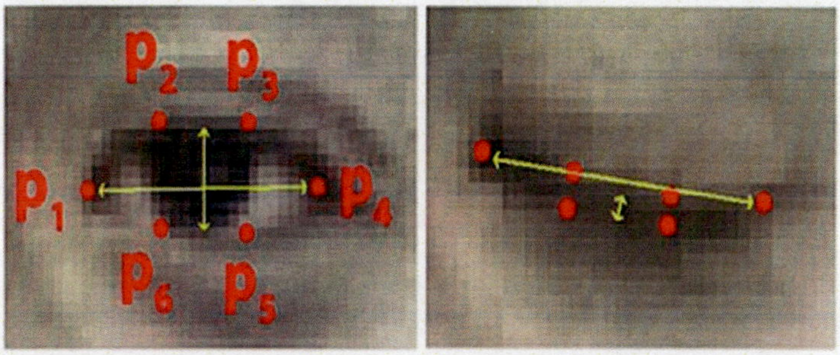


**Fig. 5.4** Landmarks for the calculation of eye aspect ratio (EAR)

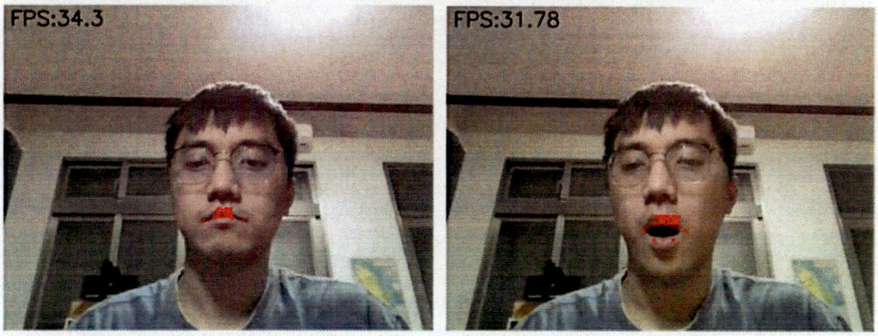


**Fig. 5.5** Use of mouth aspect ratio (MAR) to determine whether the mouth is closed or open

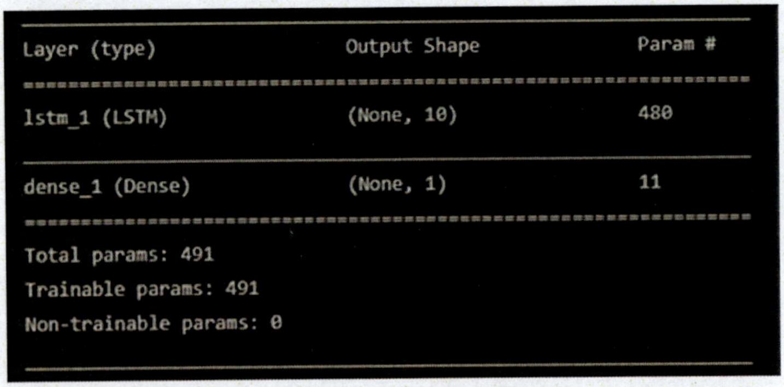

```
Layer (type)                 Output Shape              Param #
=================================================================
lstm_1 (LSTM)                (None, 10)                480
_________________________________________________________________
dense_1 (Dense)              (None, 1)                 11
=================================================================
Total params: 491
Trainable params: 491
Non-trainable params: 0
_________________________________________________________________
```

**Fig. 5.6** Parameters of our long short-term memory (LSTM) model

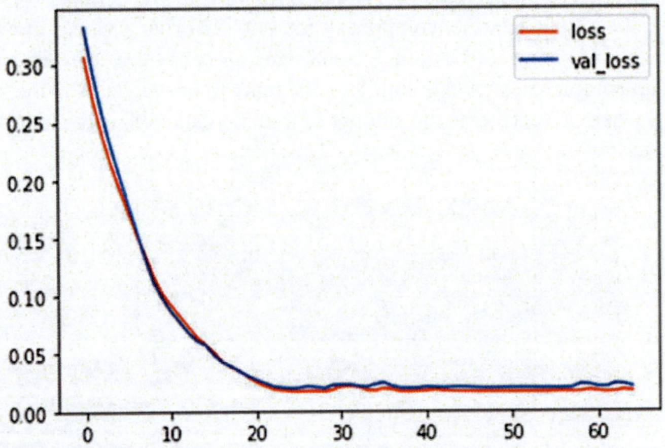


**Fig. 5.7** Loss function versus epochs

each training data is composed of, and 1 is the vector dimension in each time unit. Since we are only interested in the mouth-closing behavior, the vector dimension is 1. About 200,000 training data are obtained through the preprocessing of the dataset. Figure 5.7 shows the loss function of training (red) and validation (blue) against epochs. Figure 5.8 shows the result of yawning levels at different time steps. The curve labeled x_test corresponds to the value of MAR obtained by Eq. (5.1), that labeled y_test represents the actual labeled results in the training dataset, and that labeled predicted reflects the result after the neural network model has been trained using the dataset. The result shows that the model prediction is very close to the actual situation.

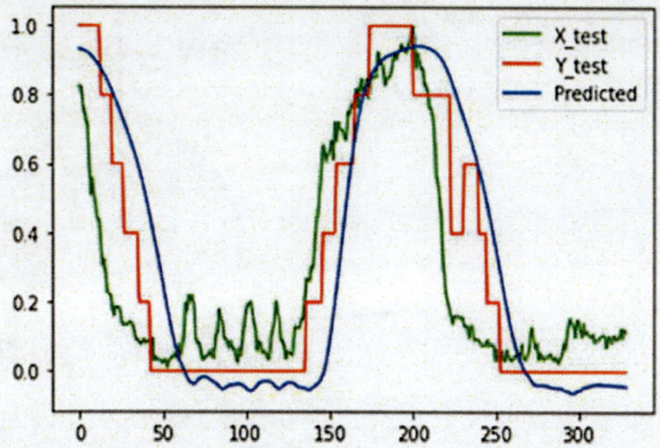


**Fig. 5.8** Yawning level at different time steps

## 5.3.3 *Facial Rotation Analysis*

BlazeFace is used to mark 468 3D features on the face. Here, four of these feature points are used to obtain the degree of rotation (roll, yaw, pitch) of the face, as illustrated in Fig. 5.9. The figure superposes the feature points on the face model in [24]. The coordinates of a point along the three axes denote by $x$, $y$, and $z$. The calculations are shown in Eqs. (5.2) through (5.4), where $c$ is a scaling parameter that needs to be calibrated once at the beginning, and the subscripts are the numbers of the feature points in the figure.

$$\text{Roll} = \tan^{-1}\left(\frac{y_{446} - y_{226}}{x_{446} - x_{226}}\right) \tag{5.2}$$

$$\text{Yaw} = (z_{446} - z_{226})\,c \tag{5.3}$$

$$\text{Pitch} = (z_{10} - z_{164})\,c. \tag{5.4}$$

Figure 5.10 shows the results when the face is rotated in different directions.

## 5.3.4 *Facial Recognition*

When face classification is used to distinguish different persons (e.g., different drivers of a bus), facial features need to be recorded by a camera. The recording

Fig. 5.9 Feature points used for determining the head rotation

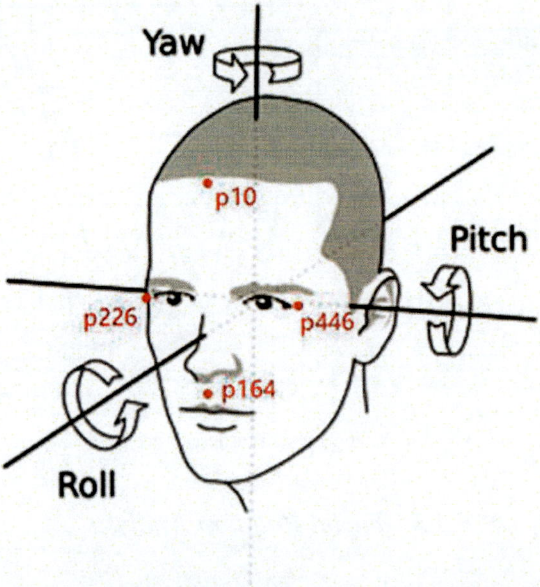


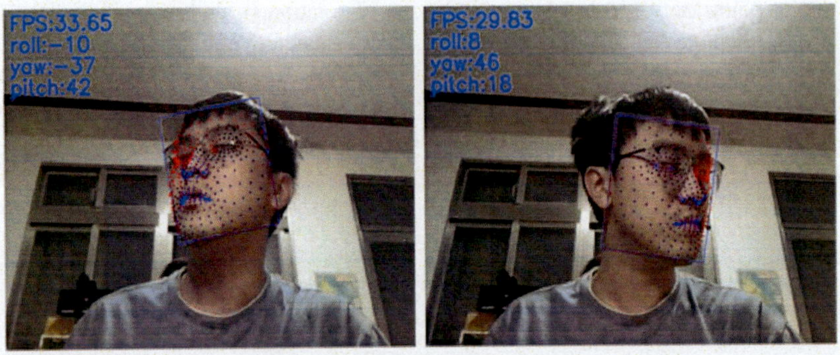


Fig. 5.10 Values of roll, yaw, and pitch for different head movements

requires an individual to face the camera and turn the head slightly so that each feature of the face is clearly recorded. The overall recording takes about 10 s. Each frame of the recorded video will be captured, and the features will be stored with the person's name tag. After that, users with features stored can be accurately identified through a camera.

In order to improve the speed and accuracy of face classification, we use the face classification method in three steps. The first step is to retrieve the ROI (regions of interest) of the face by using the BlazeFace facial recognition model. The reason for using the BlazeFace model is its speed, high accuracy, and multiple feature information that can be shared with other methods to reduce the computation. The

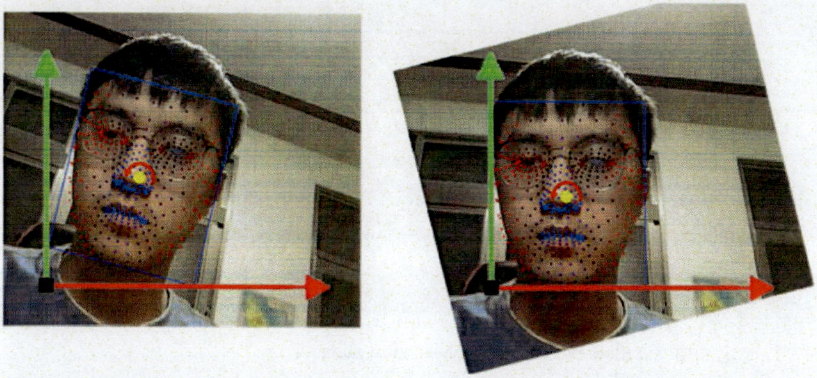

Fig. 5.11 Face alignment

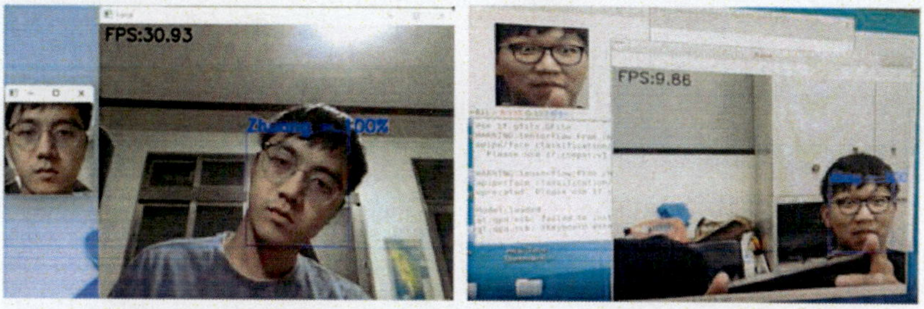


Fig. 5.12 Distinguishing the faces of two individuals

second step is to rotate the ROI for face alignment. This is accomplished by rotating the face frame and determining the pivot as the intersection of the two diagonals. The calculation is accelerated using the OpenCV matrix functions. After the rotation, the coordinates of the four corners are updated. The above steps are illustrated in Fig. 5.11. Finally, the Facenet model is used to classify different faces and output the results. Figure 5.12 shows the results of distinguishing two individuals.

### 5.3.5 Information Entropy of Facial Motion for Data Reduction

Entropy is the average amount of information contained in each message received, also known as information entropy. Entropy represents the uncertainty of a message, and a higher entropy indicates a more random overall message. The facial motion entropy can be calculated using the method proposed in [23], which is based on the distribution of the distances of the feature points from the center of the face.

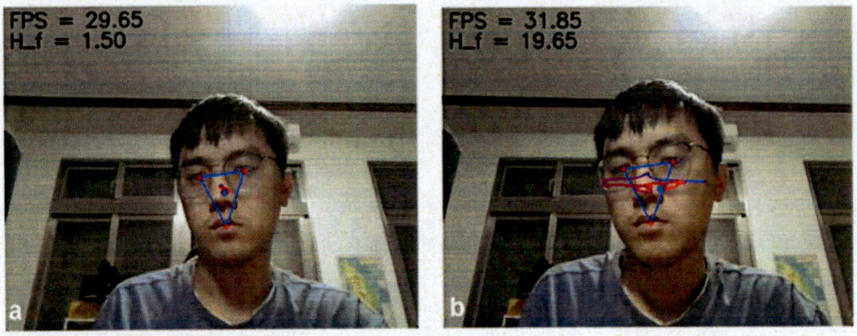


**Fig. 5.13** (**a**) Slow and (**b**) intense head movement

**Table 5.2** Training data extracted from the YawDD dataset

| Tag file information | | | Image feature extraction | | | | | | |
|---|---|---|---|---|---|---|---|---|---|
| No. | Driving status | Yawning level | Mouth closure | Eye closure | | Facial motion information entropy | Facial triaxial rotation data | | |
| Index | Mode | Level | MAR | EAR_l | EAR_r | Entropy | Roll | Yaw | Pitch |

Using the information entropy of facial motion, we can determine whether the facial movement is random or regular over a period of time, as demonstrated in Fig. 5.13.

## 5.4 Machine Learning Models

We used the YawDD dataset [25], which contains 322 videos of drivers yawning, talking, singing, etc. The facial landmark data contains the three states (normal, talking, and fatigued) and the level of yawning of a driver in a video. Then, we generated a total of more than 200,000 training data by the feature extraction method mentioned above. The training data contains the information shown in Table 5.2.

After several model training tests, we finally used the training data shown in Table 5.3. In order to reduce the input data dimension of the model, we exploit physical conditions. For example, the eye closure is almost the same for both eyes, so the X data (input data) only need to consider a single-eye closure. In addition, the inclusion of facial triaxial rotation data would greatly increase the computation of the model. Therefore, we use the facial motion information entropy instead to reduce the computational effort significantly.

In Table 5.3, X data stands for the input and Y data for output. The single eye closure is determined as follows:

$$\text{Single eye closure} = \begin{cases} \text{EAR if yaw} < 0 \\ \text{EAR if yaw} \geq 0 \end{cases} \tag{5.5}$$

**Table 5.3** Simplified dataset used in the test

| X data | | | Y data |
|---|---|---|---|
| Mouth closure | Single eye closure | Facial motion information entropy | Driving status |
| Duration of 300 frames, totally 300 × 2 data | | Single value | 0 ~ 1 representing fatigue level |

Two machine learning models are proposed. The first fully connected model is shown in Fig. 5.14. The sequence data input is first obtained by 1D convolutional layer, and the amount of data is reduced by increasing the convolutional kernel stride and the number of pooling layers. The high-dimensional sequence data is then flattened into a single dimension by the flatten layer, and the input data of facial motion information entropy is combined by the concatenate layer. Finally, the result is output by the fully connected layer (dense_22 in Fig. 5.14) with a total of 1,813,057 parameters.

Since the high-dimensional data generates a large number of parameters when the flatten layer is followed by the fully connected layers, we revised our model and use LSTM layer instead of the flatten and fully connected layers. This greatly reduces the number of parameters and the overall model size. Moreover, the LSTM has a better training effect for time-series data. The overall model design is shown in Fig. 5.15. The total number of parameters is reduced to 38,017, which is 47 times smaller than the fully connected model.

The training results of the two models are shown in the two confusion matrices in Fig. 5.16, where 0 means no fatigue is detected and 1 means fatigue is detected. The accuracy of the fully connected model is 98.95%, and that of the LSTM model is 98.73%. Thus, the accuracy of the two models is comparable, but the overall size of the LSTM model is much smaller and, therefore, more suitable for edge devices. Note that in [19], the accuracy is 94.32% using the same YawDD dataset, which is lower than our proposed models.

Even though the models have achieved high accuracy, careful analysis suggests that they might be slightly vulnerable to overfitting. The problem can be attributed to the insufficient amount of data in the training set. We have searched several datasets and referred to papers on fatigue driving. It was found that the datasets available for the training and testing of models on fatigue driving are quite scarce. The YawDD dataset used in our study is the more popular dataset at present. However, the amount of data is still not large enough to avoid overfitting. The overfitting issue may lead to lower accuracy in practical applications.

## 5.5 Conclusion

Fatigue driving can have devastating consequences for individuals and families. Body and property losses can be substantial. In this paper, we described novel

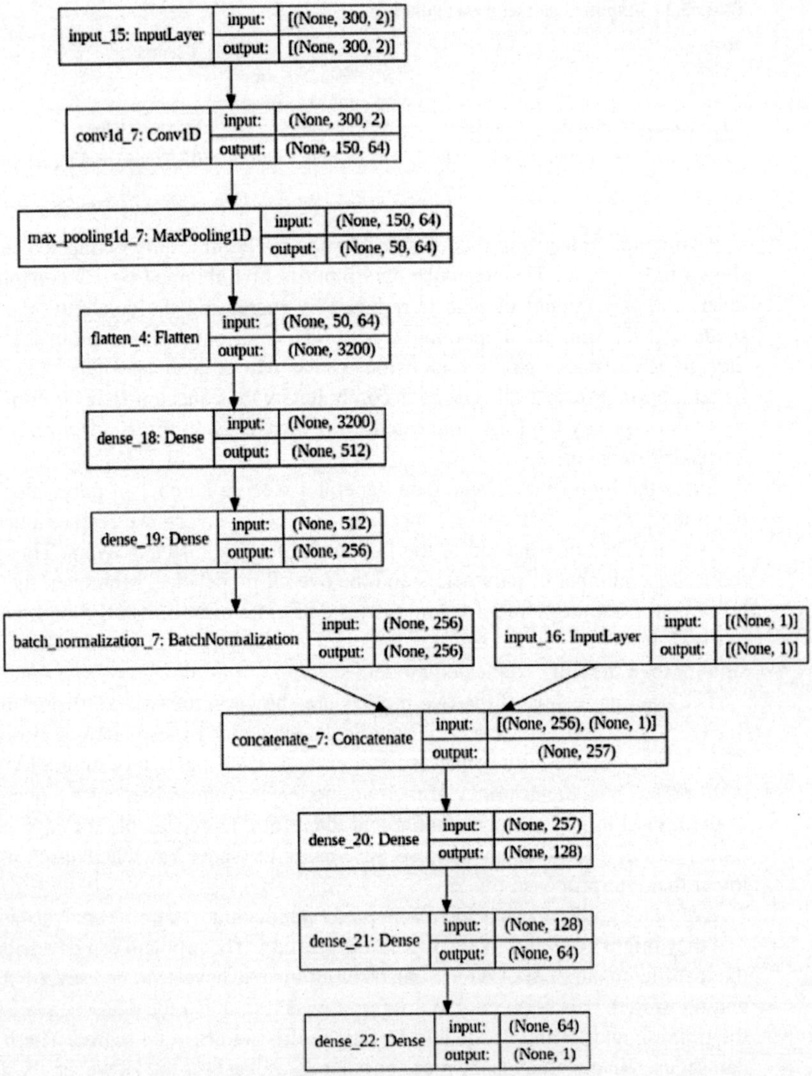


**Fig. 5.14** Fully connected machine learning model

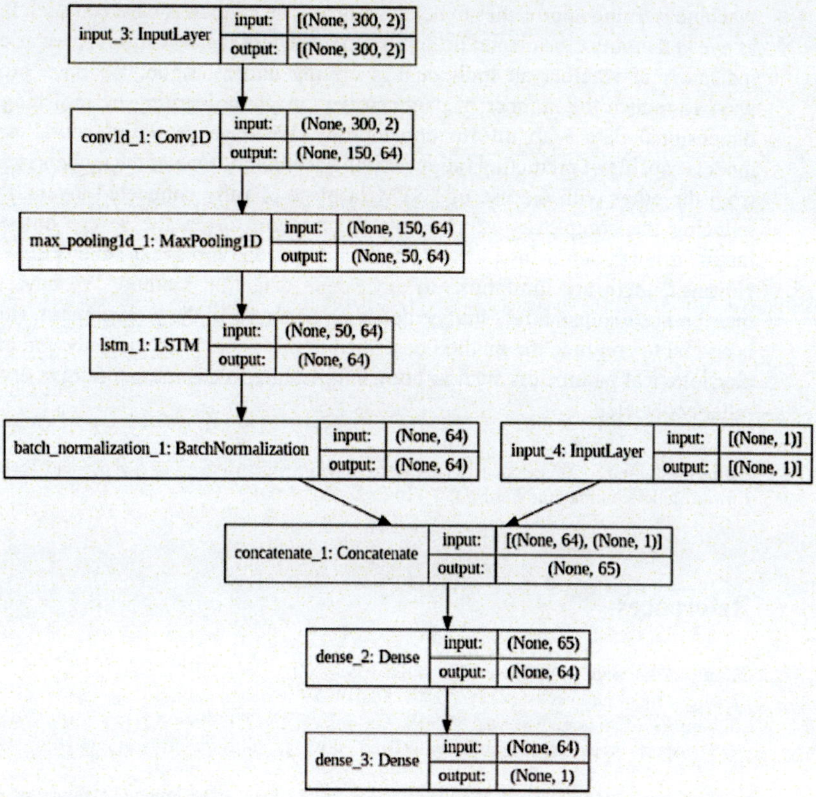


**Fig. 5.15** The machine learning model with LSTM

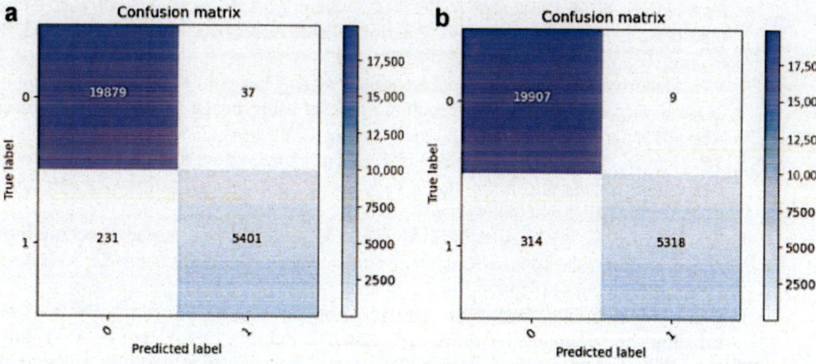


**Fig. 5.16** Confusion matrix for (**a**) the fully connected model and (**b**) the long short-term memory (LSTM) model

machine learning approaches to detect fatigue. The method is based on such features as eye and mouth closure and head movement. These facial features serve as critical indicators of whether an individual is driving under fatigue. We have proposed ways to reduce the number of parameters in machine learning by replacing high-dimensional data with information entropy. We have proposed neural network models capable of predicting fatigue driving accurately. One of the models is derived from the other with the use of LSTM in place of fully connected layers, thereby reducing the complexity of the overall structure. Given the severe outcome of fatigue driving, achieving as high a correct detection ratio as possible is imperative. However, there are limitations to any single detection method. We have shown interference that can affect fatigue detection accuracy in the environment. Hence, it is crucial to combine the method described in this paper with the detection of other physiological parameters such as body temperature to ensure that fatigue driving is reliably detected.

**Acknowledgments** This work was supported in part by the Ministry of Science and Technology, Taiwan, under contract number MOST 110-2622-E-197-001.

## References

1. https://stat.motc.gov.tw/mocdb/stmain.jsp?sys=100
2. https://money.udn.com/money/story/8888/4197576
3. https://auto.ltn.com.tw/news/9892/43
4. Transport Accident Commission: Tired driving, https://www.tac.vic.gov.au/road-safety/staying-safe/tired-driving
5. M. Ramzan, H.U. Khan, S.M. Awan, A. Ismail, M. Ilyas, A. Mahmood, A survey on state-of-the-art drowsiness detection techniques. IEEE Access **7**, 61904–61919 (2019). https://doi.org/10.1109/ACCESS.2019.2914373
6. R. Alharbey, M.M. Dessouky, A. Sedik, A.I. Siam, M.A. Elaskily, Fatigue state detection for tired persons in presence of driving periods. IEEE Access **10**, 79403–79418 (2022). https://doi.org/10.1109/ACCESS.2022.3185251
7. V.U. Maheswari, R. Aluvalu, M.P. Kantipudi, K.K. Chennam, K. Kotecha, J.R. Saini, Driver drowsiness prediction based on multiple aspects using image processing techniques. IEEE Access **10**, 54980–54990 (2022). https://doi.org/10.1109/ACCESS.2022.3176451
8. T.P. Nguyen, M.T. Chew, S. Demidenko, Eye tracking system to detect driver drowsiness, in *Proceedings of the 6th International Conference on Automation, Robotics and Applications (ICARA)*, (IEEE, 2015), pp. 472–477
9. H.S. AlZu'bi, W. Al-Nuaimy, N.S. Al-Zubi, EEG-based driver fatigue detection, in *Proceedings of the 6th International Conference on Developments in eSystems Engineering (DESE)*, (IEEE, 2013), pp. 111–114
10. G. Li, W.-Y. Chung, Detection of driver drowsiness using wavelet analysis of heart rate variability and a support vector machine classifier. Sensors **13**(12), 16494–16511 (2013)
11. H.A. Rahim, A. Dalimi, H. Jaafar, Detecting drowsy driver using pulse sensor. J. Technol. **73**(3), 5–8 (2015)
12. M. Awais, N. Badruddin, M. Drieberg, A hybrid approach to detect driver drowsiness utilizing physiological signals to improve system performance and wearability. Sensors **17**(9), 1991 (2017)

13. N. Dalal, B. Triggs, Histograms of oriented gradients for human detection, in *IEEE Computer Society Conference on Computer Vision and Pattern Recognition (CVPR '05)*, vol. 1, (IEEE, 2005), pp. 886–893. https://doi.org/10.1109/CVPR.2005.177
14. M.H. Chung, Pedestrian detection technique based on HOG algorithm and SVM classifier. M.S. Thesis, Department of Electrical Engineering, Southern Taiwan University of Science and Technology (2017)
15. K. Zhang, Z. Zhang, Z. Li, Y. Qiao, Joint face detection and alignment using multitask cascaded convolutional networks. IEEE Signal Process. Lett. **23**(10), 1499–1503 (2016). https://doi.org/10.1109/LSP.2016.2603342
16. V. Bazarevsky, Y. Kartynnik, A. Vakunov, K. Raveendran, M. Grundmann, BlazeFace: Sub-millisecond neural face detection on Mobile GPUs. ArXiv. https://arxiv.org/abs/1907.05047
17. F. Schroff, D. Kalenichenko, J. Philbin, FaceNet: A unified embedding for face recognition and clustering. ArXiv. https://arxiv.org/abs/1503.03832
18. K. He, X. Zhang, S. Ren, J. Sun, Deep residual learning for image recognition. ArXiv. https://arxiv.org/abs/1512.03385
19. F. You, Y. Gong, H. Tu, J. Liang, H. Wang, A fatigue driving detection algorithm based on facial motion information entropy. J. Adv. Transp. **2020**, 1–17, Article ID 8851485 (2020)
20. A. John, B. Cardiff, D. John, A 1D-CNN based deep learning technique for sleep apnea detection in IoT sensors. ArXiv. https://arxiv.org/abs/2105.00528
21. L. Chen, G. Xin, Y. Liu, J. Huan, Driver fatigue detection based on facial key points and LSTM. Secur. Commun. Netw. **2021**, 1–9, Article ID 5383573 (2021). https://doi.org/10.1155/2021/5383573
22. W. Zaremba, I. Sutskever, O. Vinyals, Recurrent neural network regularization. ArXiv. https://arxiv.org/abs/1409.2329
23. T. Soukupova, J. Cech, Real-time eye blink detection using facial landmarks. Computer Vision Winter Workshop, Rimske Toplice, Slovenia, 3–5 February 2016, http://vision.fe.uni-lj.si/cvww2016/proceedings/papers/05.pdf
24. T. Jantunen, J. Mesch, A. Puupponen, J. Laaksonen, On the rhythm of head movements in Finnish and Swedish Sign Language sentences, in *The Proceedings of Speech Prosody*, vol. 8, (The International Speech Communication Association (ISCA), 2016), pp. 850–853
25. S. Abtahi, M. Omidyeganeh, S. Shirmohammadi, B. Hariri, YawDD: A yawning detection dataset, in *Proceedings of the ACM Multimedia Systems*, (ACM, 2014), pp. 24–28

# Chapter 6
# Trust-Based Mechanism for Secure Communication in Fog-Based IoT

**Satish Kumar Singh and Sanjay Kumar Dhurandher**

## 6.1 Introduction

IoT is a network of interconnected devices linked to the internet to send and receive the information among these devices. IoT is a scalable network that is always growing and learning in nature. This network is not restricted to desktop computers, tablets, smartphones, and laptops only. An IoT network also includes devices like sensors, actuators, devices with increased processing power, services, and objects. Now with IoT, practically all devices can be linked to the internet and surveil remotely also. IoT is helping to ease the way humans live their lives. For example, if someone has left from his home to his office and forgot to switch off the air conditioner (AC), with IoT, the person can see the status of the AC using the app in his mobile and even turn it off by touching on his mobile. This can save the person's run back to home. This is a practical example of IoT easing human life. With the extremely faster response, monitoring, and analytical capabilities, The IoT has been adopted in various domains and in almost all industries. IoT is being used astronomically to reduce the burden on human beings. IoT devices are present everywhere and have a huge spectrum of applications like smart cities, smart homes, healthcare, and smart grids. The future of IoT seems more encouraging and propitious than ever before.

According to experts, there will be over 100 billion devices by the year 2025 [1]. IoT is an always growing and learning network. The integration of IoT and other technologies like artificial intelligence, machine learning and cloud computing is paving the way for many new and exciting innovations.

S. K. Singh (✉) · S. K. Dhurandher
Netaji Subhas University of Technology, New Delhi, India



I. Woungang, S. K. Dhurandher (eds.), *5th International Conference on Wireless, Intelligent and Distributed Environment for Communication*, Lecture Notes on Data Engineering and Communications Technologies 174,
https://doi.org/10.1007/978-3-031-33242-5_6

With an aggressive rise in the number of IoT devices, the security of the IoT network has become a major concern. Cloud computing technology used to be the solution provider for the fine-grained real-time problems in the communication networks. However, still the latency, quick time response, congestion in the network, the overall communication cost, and different types of security attacks were the major issues. Fog computing as an enhancement in the cloud computing paradigm has brought the processing near to the end devices. Fog computing is truly applicable in various domains like wireless sensors networks, smart grids, connected vehicles, software-defined networks, smart traffic light system, and IoT networks, and it has overcome these issues tremendously. The IoT network size is growing and so are the security issues, as the fog computing is still in infant stage and needs these security issues to be dealt with seriously.

The organization of the rest of this paper is as follows. The literature review with observations is presented in Sect. 6.2. Section 6.3 discusses the system model and the motivation behind this work with the trust classification model and the proposed algorithm. Simulation setup and Results are analyzed in Sect. 6.4, and finally Sect. 6.5 concludes this paper with the Conclusion of the work.

## 6.2 Literature Review

The contemporary findings of fog computing security and privacy were discussed by Alzoubi et al. [2]. They listed new provocations in security and privacy of fog computing. Further, their findings revealed that fog computing is the extension of cloud computing, and it is observed by the researchers that it is still in infant stage of development. Many queries are still unanswered in security concerns of fog computing.

Many works have been published in the domain of fog computing security and privacy. Various trust computing mechanisms for enhancing security are proposed. Some of the interesting works are discussed below:

Kandhoul et al. [3] worked on opportunistic IoT and proposed a trust-based routing protocol that calculates the trust value of a node as a weighted average of the direct and indirect trust. This protocol provides safeguard to a network against attacks like Sybil attack, packet fabrication attack, black-hole attack, good and bad-mouthing attack. Their T_CAFE protocol improves the security of opportunistic networks.

Rathee et al. [4] have casted down the various communication concerns of fog nodes and end device layer attacks during the process of handoff in fog-based networks. The malicious fog nodes and end devices are detected using the trust manager fixed between fog layer and end-user layer. Further, they compute the rating of the fog node and IoT devices services requester to route the service though a highly trusted path. They compared their proposed trusted model with the traditional approach and have successfully verified it against various networking parameters.

An evaluation system was proposed by Hussain et al. [5] based on context and reputation to compute the trustworthiness of the users in the IoT environment based on fog computing. A feedback crawler system was also implemented that helped in evaluating the trust in an unbiased manner in fog networks. The approach was proved to be effective and reliable for calculating user trustworthiness.

Yuan et al. [6] have devised a lightweight trust-computing mechanism for IoT networks. This technique was based on multisource feedback information blend. Their global trust calculation scheme was successful and reliable against bad-mouthing attacks. They also proposed an algorithm to compute feedback information based on objective information entropy theory.

This mechanism was successful to overcome conventional trust schemes.

Tu et al. [7] propose a physical layer security scheme that uses the radio channel properties between IoT devices and fog nodes to detect impersonation attacks in IoT networks. They have quantified a Q-learning approach to achieve the threshold level in impersonation attack in a dynamic environment. Their method outperforms the zero-sum game and concluded that with the rise in the number of nodes, the Q-learning based accuracy of the receiver to detect benign and malicious nodes gets higher.

Wang et al. [8] have worked on enhancement of security of fog environment by proposing an authentication scheme where communicating devices share IDs and random numbers for legitimate registration of themselves in the network. The policy of using random numbers to dynamically generate session keys proved to be more secure and a better fit to the requirements of fog-computing environments with reduced overhead.

Sharma et al. [9] proposed an anomaly detection framework to secure the IoT network against DDoS attack. They proposed a CRPS metric as a single variable algorithm. The algorithm successfully identified the attack.

A trust management system (TMS) based on a two-way subjective logic is designed by Alemneh et al. [10]. This two-way trust evaluation system computes trustworthiness from end device to fog node and vice versa using various trust metrics like friendship, ownership, honesty, latency, and packet delivery ratio. Further, they use the concept of discounting and consensus operators to calculate the direct and indirect trust value.

The attacks addressed were bad-mouthing attack, self-promotion attack, opportunistic-service attack, and on–off Attack. The TMS was proved to be safe against trust-based attacks with high accuracy.

A hypergraph clustering model established on Apriori algorithm for association study in DDoS attacks in Fog Computing Intrusion Detection System was proposed by An et al. [11]. They did simulation in MATLAB and concluded that Intrusion Detection System is an effectual technique to ensure security in fog computing.

Abdali et al. [12] discussed the advancement in fog computing environment in the recent past. They talked about the concept, architecture of fog computing, its applications, advantages, and open issues. Table 6.1 shows the state-of-the-art and their drawbacks in the flog-based IoT.

**Table 6.1** The state-of-the-art and their drawbacks in the fog-based IoT

| References | Features | Software/simulator used | Attack addressed | Drawbacks/limitations/ observations |
|---|---|---|---|---|
| Kandhoul et al. [3] | Presented a trust-based scheme called T_CAFE for opportunistic IoT | Opportunistic network environment simulator | Good and bad mouthing, Sybil, black hole, and packet fabrication | Enhanced network performance |
| Rathee et al. [4] | A secure routed and handoff mechanism proposed | NS2, Virtual machine on Microsoft Azure Cloud DS2 | – | Improved network performance through highly trusted paths<br>No attack is addressed |
| Hussain et al. [5] | Discussed about a context-aware multisource trust and reputation model | Not mentioned | – | Approach is reliable to calculate the dependability of a user<br>No attack is addressed |
| Yuan et al. [6] | Devised a trusted scheme for IoT end devices based on feedback information fusion | NetLogo Event Simulator, Java | Bad mouthing | Traditional trust schemes deficiencies are overcome by proposed trust calculation scheme |
| Tu et al. [7] | Proposed a novel technique for Security at physical layer in fog network to overcome impersonation attack | Not mentioned | Impersonation | The proposed scheme guarantees to discover the impersonation attack |
| Wang et al. [8] | A secure lightweight authentication scheme for fog computing architecture was designed | AVISPA | – | The proposed protocol authenticates devices using dynamic key generation<br>The method proved to be lightweight and secure |

| | | | | |
|---|---|---|---|---|
| Sharma et al. [9] | An anomaly detection framework was proposed to block DDoS attack in fog network | Wireshark | DDoS | The TCP-SYN and ICMP attacks were detected in dataset, and authors were able to capture detailed information from the network as their model seems more robust |
| Alemneh et al. [10] | Proposed a two-way trust management system for fog computing | Java-based simulation tool | Bad-mouthing, self-promotion, opportunistic-service, and on–off attack | Trust management system has high precision and preventiveness to trust-based attacks |
| An et al. [11] | A hypergraph clustering model is proposed based on Apriori algorithm for analysis in DDoS attacks in fog-based IDS | MATLAB | DDoS attacks | Intrusion Detection System came out as an effective tool to justify security in fog computing |

## 6.3 System Model

### *6.3.1 Motivation*

The behavior of IoT devices and fog nodes in fog-based IoT networks is dynamic in nature, and mere authentication and presence of a node in the network without participation is not enough. If a node is present in a network, then it should either use or provide services to the network in one or another way. The use of permanent trust calculation is also not right due to dynamic leave and join of nodes. Therefore, there is a need for a trust-based mechanism that can overcome these issues of IoT networks. These limitations have prompted us to design a trust-based security mechanism that calculates each node's trust based on the task performed by that particular node in the network. The mechanism is based on trust calculation that includes measurement of trust in terms of Task Frequency, Task Durability, and Task Recency.

### *6.3.2 Proposed Model*

The proposed framework as shown in Fig. 6.1 has a three-layer architecture where the cloud layer is at the top. Next, we have a fog layer whose task is to provide the services to end devices and send the data to the cloud for storage purposes and at bottom the end device layer exists whose task is to request services from the fog layer. The model contains the components as described below.

- *End Device*: A node that sends a request to the broker to perform the task.
- *Broker Node*: A broker node sends an acknowledgment to the end device that it has received a request and asks the neighboring fog node to perform. If the acknowledgment is not received or the task is not assigned to the requested fog node then the broker asks the next fog node.
- *Fog Node*: A fog node accepts the request form the broker node for performing tasks subject to its availability for processing. When no fog node is available for processing, the request is dropped.
- *Malicious Node*: Any node that hinders the performance of the networks by any means is said to be a malicious node.

The authentication process is done for each and every node in the network before starting any kind of communication. Authentication is done separately for end device, fog node, and broker. Once the authentication is done, trust is computed for nodes. Figure 6.2 shows the authentication process between access point and end device. An end device need s to associate itself with the access point in the network before any communication. Next, authentication of fog node and broker is shown in Fig. 6.3.

After this, authentication process of end device and broker is depicted in Fig. 6.4.

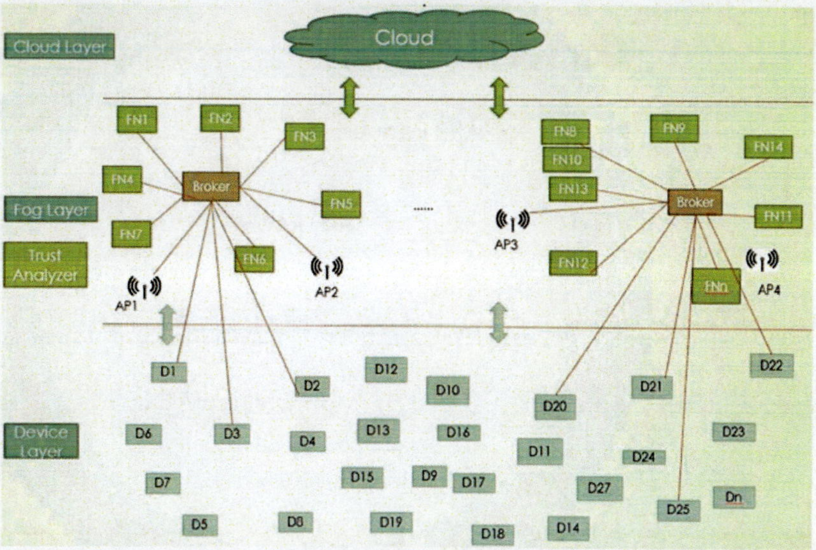


Fig. 6.1 Proposed framework

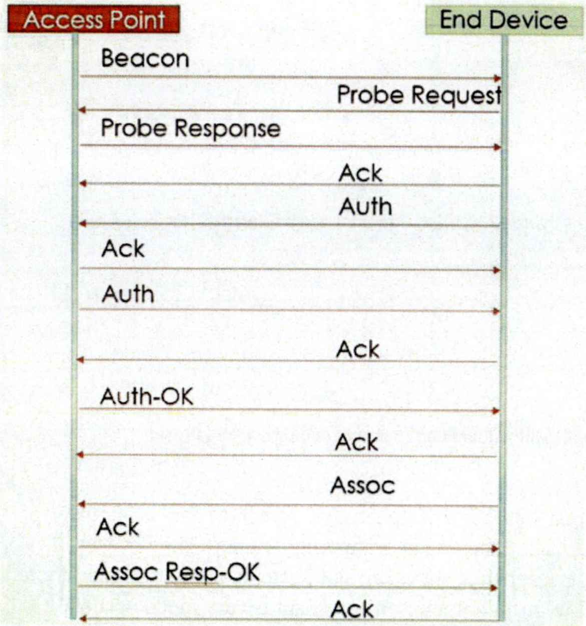


Fig. 6.2 Authentication process of access point and end device

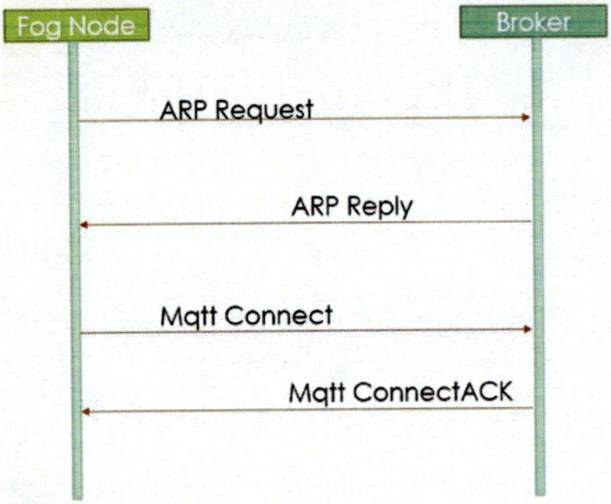


**Fig. 6.3** Authentication process of fog node and broker

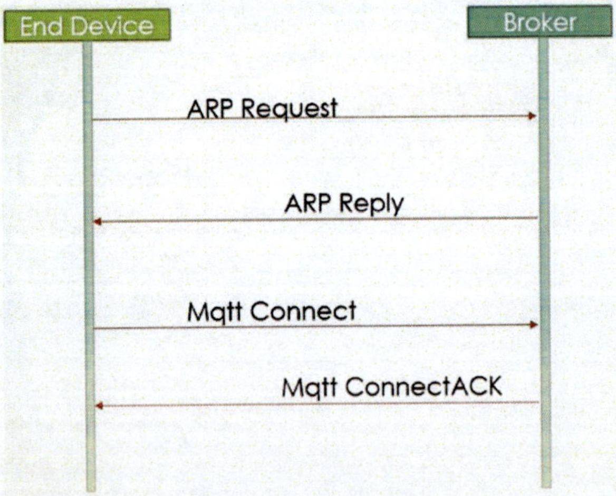


**Fig. 6.4** Authentication process of end device and broker

Trust has become the major factor as far as security of the IoT network is concerned. Many trust-computing frameworks have been proposed in the recent past, but most of them are not systematically designed therefore an efficient and reliable trust computing mechanism is very much needed.

A trusted party is always good for communication whether it is human-to-human, Machine-to-Machine or Human-to-Machine. A trusted party can always be useful to process, store and communicate in the Network in any manner. So, in Machine-

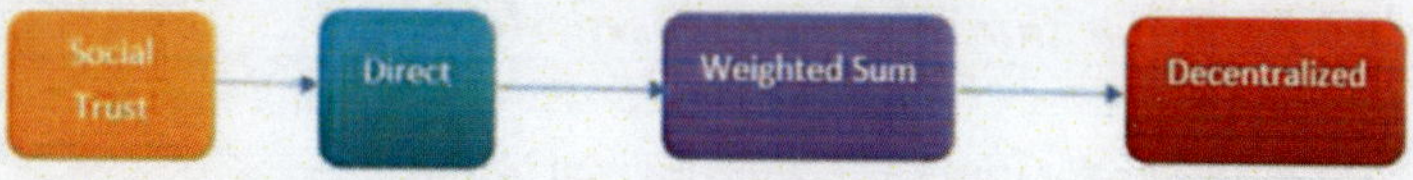


**Fig. 6.5** Trust Classification Model

to-Machine Communication in IoT networks the End Devices and Fog Devices are required to have a certain trust level to be the part of that network. Trust based techniques improves the security without lowering the communication overhead in comparison to other techniques.

Trust has the following characteristics:

- *Subjective* – It depends on trustors' perspective.
- *Asymmetric* – It does not apply in both directions between trustor and trustee.
- *Dynamic* – Trust is applicable only in a specific period.
- *Context Dependent* – The level of trust in a different view of oneself may differ remarkably.
- *Transitive* – In a situation where device A trust device B and device B trust device C, then device A may trust device C.

**Trust Classification Model**

Figure 6.5 describes m trust classification model. The trust can be measured either from QoS or from social parameters. We take social trust into consideration. The trust source can be direct, indirect, and hybrid. Here, the trust is direct in our model, and we calculate task-based trust.

The end device sends a request to the broker for task processing. Brokers in turn ask from fog node to process and complete the request in a particular time frame. Fog node sends acknowledgment of task assignment, and broker assigns the task to the fog node. A fog node, in turn, completes the task and sends the task completion message back to the broker, and broker finally sends it back to the user.

Once the task is completed, the trust algorithm uses various task-related parameters with weighted sum to calculate the direct trust of that node. If the calculated trust value falls beyond the considerable threshold, then that particular node is mentioned as a malicious node, and we stop this node from further communication in the network. The architecture for trust calculation is decentralized in our network as the trust value is not dependable on a single node.

This trust can be calculated using many factors. Here in this paper, we are using the task-based trust of a device. Trust is calculated using Task Frequency, Task Durability, and Task Recency of a particular node in connection with the fog node. The trust values are kept at a certain threshold to measure the legitimacy of a node. If the trust value is beyond a certain threshold, then that node is not allowed to communicate in the network.

### 6.3.3 Algorithm of the Proposed Model

The algorithm for trust calculation is given below:

Step 1. Input all Fog Nodes and End Devices
Step 2. Initialize all nodes with initial trust value as *t*
Step 3. For each Node *i*
Step 4. Check if Node present in *MN* Table then Malicious node encountered
Step 5. Else Compute Task Frequency, Task Durability, Task Recency
Step 6. Compute Task Frequency as: *TFi* (No of Tasks of Node *i* completed by Fog Nodes)/(Total no of Tasks completed by Fog Nodes)
Step 7. Compute Task Durability as: *TDi* (Duration of Task completed of Node *i* by Fog Nodes)/ (Total duration of all task completed by Fog Nodes)
Step 8. Compute Task Recency as: *TRi* (Time lapsed since last task of Node *i*completed by Fog Nodes/(Total time lapsed since last Task of all nodes completed by Fog Nodes)
Step 9. Compute Direct Trust of *i* Node as: Direct Trust = *(Alpha*TFi + Beta*TDi +Gamma*TRi)* where *Alpha*, *Beta*, and *Gamma* are the constants such that *Alpha+Beta+Gamma=1*
Step 10. Check If Direct Trust Parameter value of Node beyond the Threshold, then Malicious node encountered then make entry in *MN* Table and remove node from network.
Step 11. Else Assign that Direct Trust Value as the current Direct Trust value and allow Node to communicate in the network.

## 6.4 Simulation and Analysis

The proposed mechanism is simulated using FogNetSim++ simulator. The participant nodes are wirelessly connected end devices to communicate in radio medium.

We have run five different simulations setups with 5, 10, 15, 20, and 25 end devices. The simulation parameters considered are shown in Table 6.2. Each setup is run with two fog nodes and a broker. Trust is calculated based on task completion of an end device. Trust is the combination of Task Frequency, Task Durability, and Task Recency. The parameters used for trust calculation are Task Frequency, Task Durability, Task Recency, alpha, beta, and gamma. The threshold range of direct trust in our simulation setups is taken as 0.4–0.8. Any device having direct trust value within this range is considered as legitimate device, otherwise device is considered as a malicious node. The maximum value of direct trust of a node as computed in our simulations is achieved as 1.20.

**Table 6.2** Simulation parameters

| Parameter | Value |
|---|---|
| Simulator | FogNetSim++ |
| Protocol | MQTT, UDP |
| Network Channel | Wireless |
| Queue Type | Drop Tail |
| Simulation Period | 10–15 hrs |
| Mac Layer Protocol | IEEE 802.11 |
| Simulation Area | 800 m × 800 m |
| Packet Size | 1024B |
| Initial Trust Value | $t = 0.5$ ($t$ value will vary in simulations) |
| Number of Brokers | 1 |
| Number of Fog Nodes | 2 (for each simulation setup) |
| Number of Users (EndDevices) | 5–25 |

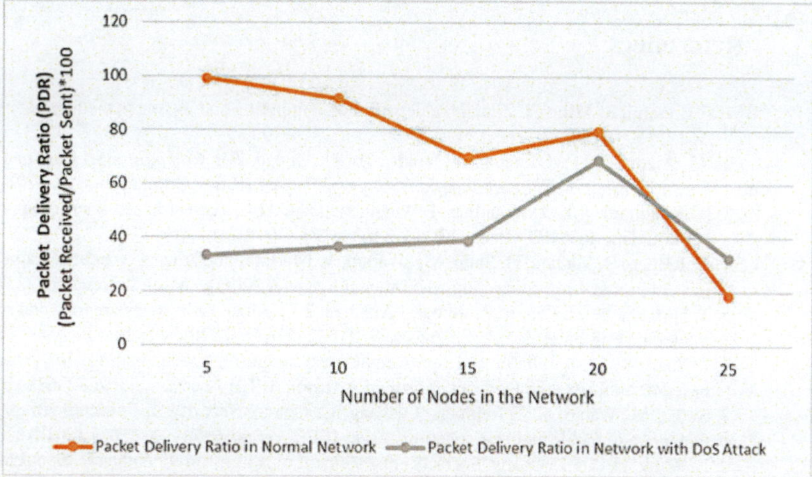


**Fig. 6.6** Packet Delivery Ratio versus Number of Nodes

## 6.4.1 Results and Analysis

The simulation of the network was done with a varying number of nodes, that is, 5–25. The network performance was analyzed with and without DoS Attack. The packet delivery ratio should remain around 80–100% in the normal network scenario. This ratio goes down in the range from 35% to 65% in the network under DoS attack. Figure 6.6 shows this downfall of PDR when a network is under DoS Attack. PDR decreases to a certain level that clearly hinders the performance of the network. We are able to successfully identify the attacker nodes in the network and remove them from the network.

## 6.5 Conclusion and Future Work

In this paper, we have proposed a trust-based security mechanism for fog-based IoT networks that calculates the direct trust of a node based on the variables like Task Frequency, Task Durability, and Task Recency with respect to the task performed by the fog node. The final trust value of a node is computed as a weighted trust of that node. After computing the trust values of all nodes, the malicious node is identified as a DoS attacker node as the node demands a huge number of tasks to be performed by the fog nodes and thereby decreasing the performance of the network in terms of packet delivery ratio. This attacker node is then removed from the network. The removal of the attacker node ensures the better performance of the network. In future, the simulations will be conducted with more nodes to identify various attacks like DDoS attack, Sybil attack, and impersonation attack in the network.

## References

1. F. Bonomi, R. Milito, J. Zhu, S. Addepalli, Fog computing and its role in the Internet of Things. ACM, 13–16 (2012)
2. Y.I. Alzoubi, V.H. Osmanaj, A. Jaradat, A. Al-Ahmad, *Fog Computing Security and Privacy for the Internet of Thing Applications: State-of-the-Art* (Wiley Online Library, 2020)
3. N. Kandhoul, S.K. Dhurandher, I. Woungang, T_CAFE: A trust based security approach for opportunistic IoT. IET Comm **13**(20), 3463–3471 (2019)
4. G. Rathee, R. Sandhu, H. Saini, M. Sivaram, V. Dhasarathan, A trust computed framework for IoT devices and fog computing environment Springer Nature. Wireless Netw **26**, 2339 (2019)
5. Y. Hussain, H. ZhiQiu, M.A. Akbar, I.A. Khan, Z.U. Khan, Context-aware trust and reputation model for Fog-based IoT. IEEE Access **8**, 31622–31632 (2020)
6. J. Yuan, X. Li, A reliable and lightweight trust computing mechanism for IoT edge devices based on multi-source feedback information fusion. IEEE Access **6**, 23626–23638 (2018)
7. S. Tu, M. Waqas, S.U. Rehman, J. Zhang, C. Chang, Security in Fog computing: A novel technique to tackle an impersonation attack. IEEE Access **6**, 74993–75001 (2018)
8. L. Wang, H. An, Z. Chang, Security enhancement on a lightweight authentication scheme with anonymity Fog computing architecture. IEEE Access **8**, 97267–97278 (2020)
9. D.K. Sharma, T. Dhankar, G. Aggarwal, S.K. Singh, J. Nebhen, I. Razzak, Anomaly detection framework to prevent DDoS attack in fog empowered IoT networks. Adhoc Netw **121**, 102603 (2021)
10. E. Alemneh, S.-M. Senouci, P. Brunet, T. Tegegne, A two-way trust management system for fog computing. FGCS **106**, 206–220 (2020)
11. X. An, J. Su, X. Lu, F. Lin, Hypergraph clustering model-based association analysis of DDOS attacks in fog computing intrusion detection system. EURASIP J. Wireless Commun. Netw Springer Open **2018**, 249 (2018)
12. Abdali T-A. N., Hassan R., . Aman MA-H., Nguyen Q. N.: Fog computing advancement concept, architecture, applications, advantages, and open issues. IEEE Access, (2021) Vol. 9. 75961–75980

# Chapter 7
# Feature Selections for Detecting Intrusions on the Internet of Medical Things

**Wei Lu, Benjamin Burnett, and Richard Phipps**

## 7.1 Introduction

The Internet of Medical Things (IoMT) is a connected networking infrastructure of medical devices, software applications, healthcare information systems, and digital health services. The connected medical devices create, collect, analyze, and then transport health data information or medical images to a cloud computing facility or internal servers using healthcare provider networks. IoMT has caused rapid changes to the healthcare industry. Medical wireless sensors, actuators, and surgical medical robots enable personalized health services and improve the quality of life for patients [1, 2]. Additionally, the health data collected by IoMT devices are stored and transmitted to advanced cloud computing platforms including IBM Watson, Amazon Web Services (AWS), and Microsoft Azure cloud services. Medical professionals can explore health-related patient data using these services to increase the accuracy of prescriptions and health-related decisions. With this capability, the IoMT has been a huge benefactor to the healthcare industry leading toward a future of smarter and more accurate diagnoses with lower costs and fewer mistakes.

However, this widespread use of IoMT devices in the medical field has given a direct pathway for many criminal activities [3]. In March 2019, the Food and Drug Administration (FDA) issued warnings about dozens of implantable cardioverter defibrillators because it was possible for them to be inflicted by malicious attacks like eavesdropping, message alteration, ransomware, and distributed denial of service [4]. This compromises patient safety, security, and the availability of critical

W. Lu (✉) · B. Burnett · R. Phipps
Department of Computer Science, Keene State College, The University System of New Hampshire, Keene, NH, USA
e-mail: wlu@usnh.edu



I. Woungang, S. K. Dhurandher (eds.), *5th International Conference on Wireless, Intelligent and Distributed Environment for Communication*, Lecture Notes on Data Engineering and Communications Technologies 174,
https://doi.org/10.1007/978-3-031-33242-5_7

healthcare systems. The current security mechanisms are poorly implemented making it easy for adversaries to hijack or remotely control a smart medical device with malware or botnet [5, 6]. With control of the medical device, they can then inject fake health data or cause malfunctions by flooding the resource-constrained IoMT with illegitimate requests. This threatens patient lives and hinders a wider deployment of IoMT devices.

To achieve the maximum possible benefit and leverage of IoMT devices in healthcare, intrusion detection techniques have been proposed to meet this urgent need to preserve the security of IoMT devices [7, 8]. Traditional intrusion detection techniques have two classification categories: misuse detection and anomaly detection [9]. Misuse detection detects attacks by matching actual behaviors recorded in audit trials with known suspicious patterns. While misuse detection is effective in uncovering known attacks, it is useless in detecting unknown or novel forms of attack where the signatures are not yet available [10–12]. Additionally, defining a signature that encompasses all variations of known attacks is a challenge [13–15]. All mistakes in the definition of these signatures increase the false alarm rate and decrease the effectiveness of the detection technique [16, 17]. Anomaly detection focuses on establishing normal activity profiles for the system [18, 19]. It assumes all intrusive activities are necessarily anomalous [20, 21]. Anomaly detection studies begin with forming an opinion on what a normal attribute for the observed object is, then decide what kinds of activities should be flagged as intrusive and how to make such decisions [22–24].

In this paper, we propose a feature selection approach to minimize a set of significant features, and then a classic logistic regression model is used to improve the general detection performance based on the selected features with significance. In terms of data collection, a publicly available dataset called WUSTL-EHMS-2020 is applied to evaluate our approach, where the initial number of raw features is 40, and after the feature selection procedure, we receive seven significant features [25, 26]. A logistic regression model for classification is then built based on these seven features for intrusion detection in the IoMT network.

This paper is organized as follows. Section 7.2 introduces the related works in the field including some of the most recent research literature to secure IoMT devices. Section 7.3 presents the descriptive statistics on the WUSTL-EHMS-2020 dataset that was collected in St. Louis by Washington University to determine the performance of machine learning algorithms for intrusion detection when attackers infiltrate the IoMT network. We analyze the raw WUSTL-EHMS-2020 dataset with some descriptive statistics so we can have a basic picture of what the data looks like and how each variable is distributed and interpreted. During this data preprocessing procedure, we eliminate some features and data points that include obvious anomalies and outliers. Section 7.4 illustrates the details of selecting significant features, and Sect. 7.5 introduces how to build a logistic regression model for classifying normal and intrusion behaviors based on the selected seven significant features. Finally, Sect. 7.6 makes some concluding remarks by comparing the accuracy, true-positive rate, and precision of our predictive model to a PCA-based anomaly detection approach and then discusses future work.

## 7.2 Related Works

There are several most recent studies working on security systems for IoMT devices. These approaches mainly include lightweight authentication or classification, access control, and encryption, considering the low number of resources that IoMT devices need to use, such as limited battery life and computational power.

In [2], Papaioannou et al. discuss several security measures against attacks on Internet of Medical Things (IoMT) edge networks. These security measures focus on attacks on confidentiality, integrity, authentication, authorization, availability, and non-repudiation. The mentioned security measures for confidentiality and integrity focus on encryption, symmetric keys, and cryptography, and the author claims this has no protection against malware. Security for authentication and authorization includes many protocols and techniques, but therein lies an issue with resource requirements as there is a limit to the resources that an IoMT device can utilize.

In [27], Wang et al. introduce an algorithm called Access-Control Determination located at the FOG server. This algorithm is designed to support decision-making when using access controls and works to lower the number of threats from over collections by cloud-based apps. In [28], Yanambaka et al. cover a method of lightweight authentication using Physical Unclonable Functions. These generate cryptography keys to check the authentication of incoming signals for medical devices. This method has no processor requirements for devices and, if using a low-power model, is a viable option for authentication with reduced power consumption. In [29], Su et al. present a lightweight classification for malware using image classification. This method converts malware binary to an 8-bit sequence and then converts that to a gray-scale image. Using machine learning, they can compare the images of malware to normal operations, which have a distinctly different look. This method predicts malware with 94% accuracy.

In [6], Sampangi et al. develop two algorithms for wireless medical device security. Independent and Adaptive Management of Keys (IAMKeys) and Key Management and Encryption for Securing Inter-Sensor Communication (KEMESIS). This uses one (KEMESIS) or two (IAMKeys) keys to encrypt data in the network. The goal is to remove security key communication and to generate the key independently by the sender and receiver so that the key cannot be compromised during communication. This method selects a random key using one of the data fields from a randomly chosen reference frame. Another key cryptosystem is presented by Sangari and Manickam in [30]. Similar to the approach in [6], this system works to remove security key communication. However, the method uses the electrocardiogram (EKG) signals from the patient to create the cryptography keys instead of random data fields.

In [31], Hady et al. present an intrusion detection system that directly connects medical devices to a sensor board, then connects that to a laptop using a USB, which acts as a gateway. The gateway uses Wi-Fi to transfer data to a server with TCP/IP protocol. The security system inside this is the use of machine learning to boost the monitoring system and detect abnormalities in real time. In [32], Gupta et al. study anomaly detection in a remote patient setting. In this research, abnormalities are defined as abnormal activities performed by the user and abnormal values from IoMT devices. They use a Hidden Markov Model (HMM) to first learn the user's normal behaviors; then, using that data, the anomaly detection method can look for any discrepancies within a certain margin.

## 7.3 Descriptive Statistics on WUSTL-EHMS-2020 Datasets

The WUSTL-EHMS-2020 dataset was created using a real-time Enhanced Healthcare Monitoring System (EHMS) testbed [25]. This testbed collects both the network flow metrics and patients' biometrics in which 32 features are collected from the network flow metrics and eight features are collected from patient vitals gathered by medical devices (EKG machines). These include temperature, SpO2 (Blood Oxygen), Pulse Rate, SYS (Systolic Blood pressure), DIA (Diastolic blood pressure), Heart Rate, Respiration Rate, and ST Segment. In the following, we analyze each variable one by one with some descriptive statistics.

*Temp* represents the temperature of patients. Healthcare professionals look at this value to see how your temperature is doing, and this also helps doctors narrow down what kind of illness you have. The attackers can change this value to mislead the doctors into giving the wrong treatment to the patient and ignoring the actual issue. As illustrated in Table 7.1, *Temp* is ranged from 23.6 to 28.9 °C over the total 14,272 normal instances with a mean of 26.92 °C. In terms of the 2046 attack instances, *Temp* is ranged from 24.1 to 29.2 °C with a mean of 26.8 °C. It is well-known that normal body temperature can range between 36.1 and 37.2 °C with an average body temperature of 37 °C. The *Temp* data must be polluted during the data collection, and thus, we will remove this feature in our following analysis.

*SpO2* represents the blood oxygen of patients. This feature is very important for healthcare professionals to look at since it tells the doctors how good your blood circulation and lungs are working. A low value of *SpO2* simply means that there is an issue with your blood circulation and lungs. The attacker can lower this value on purpose to trick the doctors into giving the wrong treatment. As illustrated in Table 7.2, *SpO2* is ranged from 0 to 100 over the total of 14,272 normal instances with a mean value of 97.84 and from 91 to 100 over the total of 2046 attack instances

**Table 7.1** Descriptive statistics on Temp

| | Feature | |
|---|---|---|
| | Temp in Celsius degrees | |
| Statistics | Normal | Attack |
| Maximum | 28.9 | 29.2 |
| Minimum | 23.6 | 24.1 |
| Mean | 26.92 | 26.8 |
| Standard deviation | 0.92 | 0.93 |

**Table 7.2** Descriptive statistics on SpO2

| | Feature | |
|---|---|---|
| | SpO2 | |
| Statistics | Normal | Attack |
| Maximum | 100.0 | 100.0 |
| Minimum | 0.0 | 91.0 |
| Mean | 97.84 | 97.6 |
| Standard deviation | 1.53 | 1.23 |

**Table 7.3** Descriptive statistics on SYS and DIA

| | Feature | | | |
|---|---|---|---|---|
| | SYS | | DIA | |
| Statistics | Normal | Attack | Attack | Attack |
| Maximum | 149.0 | 149.0 | 87.0 | 95.0 |
| Minimum | 0.0 | 129.0 | 0.0 | 72.0 |
| Mean | 142.87 | 142.72 | 80.07 | 80.29 |
| Standard deviation | 8.85 | 5.37 | 6.3 | 4.68 |

with a mean value of 97.6. Since the patients cannot have a *SpO2* value of 0.0, the data instances with a *SpO2* value of 0.0 will be removed from the dataset since we consider those data are polluted during data collection.

*Heart Rate* represents the number of heartbeats per minute, and *Pulse Rate* represents the number of pulses caused by the heart contracting. They should be essentially the same and can tell doctors how fast your heart is pumping oxygen and nutrients throughout the body. These two are very important values to look at since your heart is connected to everything. If the heart rate is increased, the heart beats too fast; hence, it may not pump enough blood to the rest of the body. As a result, the organs and tissues may not get enough oxygen. If it is slow, it will not be able to pump enough oxygen and nutrients throughout the body and will affect the patient poorly. The descriptive statistics show that *Pulse Rate* is ranged from 0 to 194 over the total 14,272 normal instances with a mean value of 76.45 and from 69 to 116 over the total of 2046 attack instances with a mean value of 78.61. Similarly, the *Heart Rate* is ranged from 0 to 119 over the total 14,272 normal instances with a mean value of 75.36 and from 44 to 110 over the total of 2046 attack instances with a mean value of 76.04. Since the patients cannot have a value of 0.0 for *Heart Rate* and *Pulse Rate*, the data instances with a value of 0.0 will be removed from the dataset since we consider those data are polluted during data collection.

*SYS* represents Systolic Blood Pressure, and *DIA* represents Diastolic Blood Pressure; they are collected using the mean of the last two of three measurements (mmHg). *DIA* and *SYS* help medical professionals determine if patients are at a high risk of stroke or heart attack. It is normal when a patient has a systolic pressure of less than 120 and a diastolic pressure of less than 80. As illustrated in Table 7.3, *SYS* is ranged from 0 to 149 over the total of 14,272 normal instances with a mean value of 142.87 and from 129 to 149 over the total of 2046 attack instances with a mean value of 142.72.

**Table 7.4** Descriptive statistics on respiration rate

| | Feature | |
|---|---|---|
| | Respiration rate | |
| Statistics | Normal | Attack |
| Maximum | 73.0 | 38.0 |
| Minimum | 0.0 | 0.0 |
| Mean | 19.71 | 19.58 |
| Standard deviation | 7.44 | 6.45 |

Similarly, *DIA* is ranged from 0 to 87 over the total 14,272 normal instances with a mean value of 80.07 and from 72 to 95 over the total of 2046 attack instances with a mean value of 80.29. Since the patients cannot have the *SYS* and *DIA* value of 0.0, the data instances with *SYS* and *DIA* value of 0.0 will be removed from the dataset since we consider those data are polluted during data collection.

*Respiration Rate* is a fundamental vital sign indicating life-threatening illnesses such as adverse cardiac events, pneumonia, clinical deterioration and stressors, emotional stress, cognitive load, heat, cold, physical effort, and exercise-induced fatigue. If attackers were to change the value of respiration rate, it would make it a lot harder for healthcare professionals to give the right treatment to the patient. As illustrated in Table 7.4, the *Respiration Rate* is ranged from 0 to 73 over the total of 14,272 normal instances with a mean value of 19.71 and from 0 to 38 over the total of 2046 attack instances with a mean value of 19.58. Since the patients cannot have the *Respiration Rate* value of 0.0, the data instances with a *Respiration Rate* value of 0.0 will be removed from the dataset since we consider those data are polluted during data collection.

*ST* is an electrically neutral segment in the heart monitor that is between ventricular depolarization (QRS complex) and repolarization (T wave) in millivolts (mv). *ST* provides important information to medical professionals since an elevation of the *ST* segment tells them whether the patient has a total blockage of the involved coronary artery and that the heart muscle is currently dying. The primary treatment for the elevation of an ST segment is called a Primary percutaneous coronary intervention (p-PCI), a procedure used to insert a camera into the patient and take a closer look at what is going on in the heart. As illustrated in Table 7.5, *ST* is ranged from −0.3 to 1 over the total of 14,272 normal instances with a mean value of 0.26 and from −0.02 to 1 over the total of 2046 attack instances with a mean value of 0.26. Since the patients cannot have the *ST*-segment value smaller than 0.0, the data instances with a negative value of *ST* segment will be removed from the dataset since we consider those data are polluted during data collection.

Using the same procedure, we conduct the descriptive statistical analysis on all of the network flow metrics, including SrcBytes, DstBytes, SrcLoad, DstLoad, SIntPkt, DIntPkt, SrcJitter, DstJitter, sMaxPktSz, dMaxPktSz, sMinPktSz, dMinPktSz, Dur, TotPkts, TotBytes, Load, and Rate. More details on these descriptive statistics can be found in [33]. After cleaning up the obvious outliers or polluted data instances of the WUSTL-EHMS-2020 dataset, we form a new dataset including 15,150 data

**Table 7.5** Descriptive statistics on ST segment

| | Feature | |
|---|---|---|
| | ST segment | |
| Statistics | Normal | Attack |
| Maximum | 1.0 | 1.0 |
| Minimum | −0.3 | −0.02 |
| Mean | 0.26 | 0.26 |
| Standard deviation | 0.1 | 0.11 |

instances in which 13,219 instances are labeled normal and 1931 instances are labeled attack.

## 7.4 Logistic Regression Model for Feature Selection

Logistic regression, or binary classification, is one of the most popular statistical methods we can use to predict a binary target variable [34]. Logistic regression models compute a value on the scale of 0 to 1, which can be interpreted as the probability of a given observation belonging to one of the two groups. This can then be used to create the classification of observation to one of the two groups of interest. In this paper, the observation of the target variable would be 0 or 1 where 0 represents the data instance is normal and 1 means the data instance is an attack. In the following, we conduct the prediction of *IntrusionLabel* and its relations with risk factors using the logistic regression model. The risk factors we use in this model include mainly the network flow metrics and biometrics of patients as described in the following:

| | |
|---|---|
| SrcBytes | Number of bytes from source to destination, denoted by $x_1$ |
| DstBytes | Number of bytes from destination to source, denoted by $x_2$ |
| SrcLoad | Source to destination bits per second, denoted by $x_3$ |
| DstLoad | Destination to source bits per second, denoted by $x_4$ |
| sIntPkt | Interpacket arrival time (millisecond) from source to destination, denoted by $x_5$ |
| dIntPkt | Interpacket arrival time (millisecond) from destination to source, denoted by $x_6$ |
| SrcJitter | Source jitter (millisecond), denoted by $x_7$ |
| DstJitter | Destination jitter (millisecond), denoted by $x_8$ |
| Dur | Transaction record total duration, denoted by $x_9$ |
| TotPkts | Total transaction packets count, denoted by $x_{10}$ |
| Load | Total transaction bits per second, denoted by $x_{11}$ |
| Rate | Number of packets per second, denoted by $x_{12}$ |
| SpO2 | Blood oxygen, denoted by $x_{13}$ |
| Pulse_Rate | Pulse Rate in BPM, denoted by $x_{14}$ |
| SYS | SYStolic blood pressure, denoted by $x_{15}$ |

| | |
|---|---|
| DIA | DIAstolic blood pressure, denoted by $x_{16}$ |
| Heart_rate | Heart Rate in Beats Per Minute (BPM), denoted by $x_{17}$ |
| Resp_Rate | Respiration Rate in BPM, denoted by $x_{18}$ |
| ST | Electrically neutral area between ventricular depolarization (QRS complex) and repolarization (T wave) in millivolts (mv), denoted by $x_{19}$ |

The predicted variable (i.e., desired target) is the risk of attacks where the value of *Label* 1 means Yes to an intrusion behavior, and the value of 0 means No to an intrusion behavior (i.e., normal).

Given the above information, we define the logistic regression model in the following:

$$\pi^* = \beta_0 + \beta_1 x_1 + \beta_2 x_2 + \cdots + \beta_{18} x_{18} + \beta_{19} x_{19} \tag{7.1}$$

where

$$\pi^* = \ln\left(\frac{\pi}{1-\pi}\right) \tag{7.2}$$

$$\pi = P\,(y = 1) \tag{7.3}$$

$$y = \begin{cases} 0 \text{ if the medical record is not changed by an attacker} \\ 1 \text{ if the medical record is changed by an attacker} \end{cases}$$

By running logistic regression, we find that the overall logistic regression model is not statistically useful for predicting the risk of attacking behaviors. This is because as illustrated in Tables 7.6 and 7.7, many of the factors have a large value of $p$-value, meaning that they are not significant in the predicting model. Therefore, we need to remove them one by one to make sure all the parameters are significant. From Table 7.6, we see *TotPkts* has the largest $p$-value of 0.8122, which is larger than the threshold $p$-value of 0.01 we set, thus we remove *TotPkts* first and then run the logistics model again. Table 7.7 shows the result of parameter estimates after removing *TotPkts*, from which we know *Rate* has the largest $p$-value of 0.6336, which is not significant; therefore, we remove *Rate* next and run the logistic model again.

Following the same steps, we run the model ten more steps and remove Dur, Resp_Rate, Scr_Bytes, SYS, Dst_Load, Load, ST, DIntPkt, Heart_Rate, and Dst_Jitter in the order because these parameters have a $p$-value of 0.9930, 0.5849, 0.3935, 0.3791, 0.2020, 0.8313, 0.0422, 0.0317, 0.0468, and 0.0390, respectively, in each run.

**Table 7.6** Parameter estimates of initial 19 features

| Term | Estimate | *p*-value | Term | Estimate | *p*-value |
|---|---|---|---|---|---|
| *SrcBytes* | −0.0098543 | 0.5752 | *Load* | 0.034808 | 0.2367 |
| *DstBytes* | 0.14874605 | 0.2268 | *Rate* | 1.12632524 | 0.1002 |
| *SrcLoad* | −0.0353931 | 0.2289 | *SpO2* | 0.17329535 | 0.0002 |
| *DstLoad* | −0.0398755 | 0.1767 | *Pulse_Rate* | 0.03279197 | <0.0001 |
| *SIntPkt* | 0.16318328 | 0.0004 | *SYS* | −0.0074055 | 0.3520 |
| *DIntPkt* | −0.0461866 | 0.3460 | *DIA* | 0.02698149 | 0.0049 |
| *SrcJitter* | −0.1859683 | <0.0001 | *Heart_rate* | 0.01168791 | 0.0338 |
| *DstJitter* | 0.0922444 | 0.0947 | *Resp_Rate* | 0.0030437 | 0.5850 |
| *Dur* | −13.019227 | 0.2951 | *ST* | 0.71037263 | 0.0299 |
| *TotPkts* | −1.0588702 | 0.8122 | | | |

**Table 7.7** Parameter estimates of 18 features

| Term | Estimate | *p*-value | Term | Estimate | *p*-value |
|---|---|---|---|---|---|
| *SrcBytes* | −0.0119989 | 0.3324 | *Load* | 0.03475589 | 0.2394 |
| *DstBytes* | 0.11847025 | 0.3131 | *Rate* | 1.16022205 | 0.6336 |
| *SrcLoad* | −0.0353738 | 0.2294 | *SpO2* | 0.17337817 | 0.0002 |
| *DstLoad* | −0.0399301 | 0.1865 | *Pulse_Rate* | 0.0327997 | <0.0001 |
| *SIntPkt* | 0.15537371 | 0.0312 | *SYS* | −0.0074135 | 0.3515 |
| *DIntPkt* | −0.043284 | 0.3857 | *DIA* | 0.02701092 | 0.0049 |
| *SrcJitter* | −0.186036 | <0.0001 | *Heart_rate* | 0.01168783 | 0.0338 |
| *DstJitter* | 0.08907791 | 0.1116 | *Resp_Rate* | 0.00304309 | 0.5851 |
| *Dur* | −10.310807 | 0.6331 | *ST* | 0.71009454 | 0.0299 |

By running logistic regression iteratively, we finally find an overall logistic model that is statistically useful for predicting the risk of attacking behaviors. As illustrated in Fig. 7.1, we have the *p*-values of SrcJitter, SIntPkt, DstBytes, Pulse_Rate, SrcLoad, and SpO2 all 0.00000, and the *p*-value of DIA is 0.0025. All of the *p*-values are smaller than the threshold *p*-value of 0.01 we set, showing that all the seven variables of risk factors are significant. Then, as illustrated in Table 7.8, we find that the chi-square value of $X^2$ is 4182.219 with a *p*-value less than 0.0001. At $\alpha = .01$, the *p*-value is less than 0.0001, which is less than $\alpha$. As a result, in this case, we have sufficient evidence to reject the null hypothesis; therefore, we can conclude that the overall logistic regression model is statistically useful for predicting the risk of attack behavior conducted in this medical network.

The logistic regression model provides a method to estimate the probability of choosing one of the response levels as a smooth function of the factor. In this paper, we use it as an approach for selecting significant features, where the detailed algorithm is illustrated in Table 7.9 using the python pseudocode.

Starting from the 19 features, the model downsizes them to 7 significant features where we can achieve a meaningful prediction regarding the target variable of intrusion behaviors. The confusion matrix also illustrates the prediction performance of

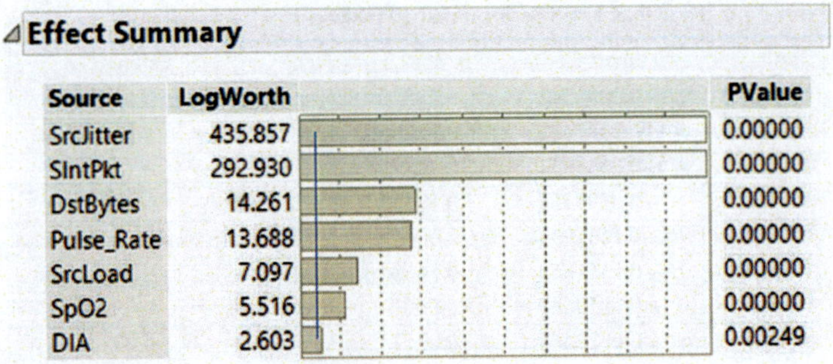


**Fig. 7.1** Effect summary on seven selected features

**Table 7.8** Whole model test on selected features

| Model | -LogLikelihood | DF | Chi-square | *p*-value |
|---|---|---|---|---|
| Difference | 2091.1094 | 7 | 4182.219 | <0.0001 |
| Full | 3689.0237 | | | |
| Reduced | 5780.1367 | | | |

**Table 7.9** Feature selection algorithm

**Function** LgitFS (data) **returns** significant features<br>
**Import** python libraries<br>
**Import** dataset<br>
data = pd.read_csv(dataset)<br>
**Create training and testing data**<br>
trainA, testA, trainB, testB =<br>
train_test_split(A, B, test_size = 0.2, random_state = 123456)<br>
**Call logit** (trainB, trainA):<br>
logit_model = Logit(trainB, trainA)<br>
result = logit_model.fit()<br>
print(result.summary())<br>
**If** (p-value>0.05) **Then** loop back to logit()<br>
**Else** break<br>
**Return** $SF_i, \quad i = 1, 2, \ldots, k$

**Table 7.10** Confusion matrix

| | Predicted | |
|---|---|---|
| | Count | |
| Actual label | 1 | 0 |
| 1 | 877 | 1054 |
| 0 | 8 | 13,211 |

our testing model. As illustrated in Table 7.10, we can find that the overall accuracy of our model is 92.99%.

Based on the table above, we can derive the following [35]:

- True positives (TP) is 877, that is, these are cases in which we predicted intrusion to the data instances, and they do are intrusions.
- True negatives (TN) is 13,211, that is, we predicted normal to the data instances, and they do are normal.
- False positives (FP) is 8, that is, we predicted intrusion to the data instances, but they do are normal.
- False negatives (FN) is 1054, that is, we predicted normal to the data instances, but they do are intrusions.

A set of metrics can be derived based on the value of TP, TN, FP, and FN [36]:

- *Accuracy*: measures how often the model is correct overall. In this case, we have:

$$\text{accuracy} = \frac{\text{TP} + \text{TN}}{\text{total}} = \frac{(877 + 13211)}{15150} = 0.9299$$

- *Misclassification Rate*: measures how often the model is wrong overall. In this case, we have the error rate (or misclassification rate) is equal to 0.0701
- *True-Positive Rate*: measures when the data instances are actually intrusions and how often our model predicts them to be intrusions. In this case, we have:

$$\text{True Positive Rate} = \frac{\text{TP}}{\text{actual intrusion}} = \frac{877}{1931} = 0.4542$$

- *False-Positive Rate*: measures when the data instances are actually normal and how often our model predicts them to intrusions. In this case, we have:

$$\text{False Positive Rate} = \frac{\text{FP}}{\text{actual normal}} = \frac{8}{13219} = 0.00061$$

- *True-Negative Rate*: measures when the data instances are actually normal and how often our model predicts them to be normal. In this case, we have:

$$\text{True Negative Rate} = \frac{\text{TN}}{\text{actual normal}} = \frac{13211}{13219} = 0.9994$$

- *False-Negative Rate*: measures when the data instances are actually intrusions and how often our model predicts them to be normal. In this case, we have:

$$\text{False Negative Rate} = \frac{\text{FN}}{\text{actual intrusion}} = \frac{1054}{1931} = 0.5458$$

- *Precision*: measures when the model predicts data instances to intrusions and how often the model is correct. In this case, we have:

$$\text{Precision} = \frac{\text{TP}}{\text{predicted intrusion}} = \frac{877}{885} = 0.991$$

- *Prevalence*: measures how often the intrusion condition actually occurs in the sample of our data instances. In this case, we have:

$$\text{Prevalence} = \frac{\text{actual intrusions}}{\text{total}} = \frac{1931}{15150} = 0.1275$$

The overall accuracy rate of our model is about 93% given the cutoff value of 0.5. This is good because when talking about this predictive model we generally want to do better than 50%. This is what we would expect to get if we were just randomly guessing a randomly assigned value of normal or intrusion outcome to each individual in this dataset, and we would get about 50% correct classification [37]. The model achieves a 93% accuracy of classification rate, which is much higher than 50% of random guessing. To evaluate the performance of our classification model further, we need to consider both the false-positive rate and the true-positive rate (i.e., Sensitivity) so we can achieve a tradeoff between these two essential performance evaluation metrics.

## 7.5 Predicting Intrusions Using Selected Features

Sensitivity is essentially the true-positive rate, and it measures when the data instances are actually intrusions and how often our model predicts them to be intrusions. In the overall classification, the higher the value of Sensitivity, the better the performance of the prediction model. Specificity, on the other hand, is the complement of the false positive rate, and it measures when the data instances are normal and how often our model predicts them to be normal. Similarly, in the overall classification, the higher the value of Specificity, the better the performance of the prediction model. In practice, we expect a high value for both Sensitivity and Specificity, or in other words, when evaluating the prediction model, we expect a high value for Sensitivity and a low value for false-positive rate simultaneously. Given the default cutoff value of 0.5, our model generates a very low value of false-positive rate, that is, 0.061%; however, its Sensitivity (i.e., the true positive rate) has a low value of 45.42% as well, simply meaning that the model fails to detect many intrusions.

To address this issue, we need to change the cutoff point for the output variable. By default, the cutoff point is 0.5, which means if the probability of the output is larger than 0.5, the classification would be an intrusion, and otherwise, if the probability is smaller than 0.5, the classification would be normal. By changing the value of the cutoff point, we can achieve an optimal cutoff point so that the model can generate optimal values for both true-positive rate and false-positive rate. The ROC (Receiver Operating Characteristic) curve can help show us the tradeoff between true positives and false positives at different cutoff values. Figure 7.2 illustrates the ROC analysis of our prediction model, where we detect an optimal

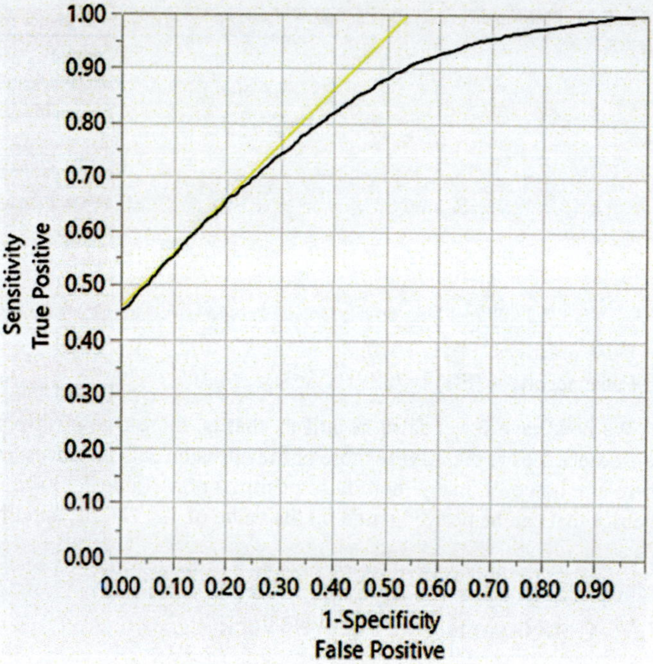


**Fig. 7.2** Receiver operating characteristic (ROC) analysis

**Table 7.11** Samples in receiver operating characteristic (ROC) table

| Probability | 1-Specificity | Sensitivity | True positive | True negative | False positive | False negative |
|---|---|---|---|---|---|---|
| 0.0901 | 0.1486 | 0.6038 | 1166 | 11,254 | 1965 | 765 |
| 0.0900 | 0.1487 | 0.6044 | 1168 | 11,249 | 1970 | 763 |
| 0.0899 | 0.1499 | 0.6054 | 1169 | 11,238 | 1981 | 762 |

**Table 7.12** Confusion matrix with cutoff point 0.09

| | Predicted | |
|---|---|---|
| | Count | |
| Actual label | 0 | 1 |
| 0 | 11,249 | 1970 |
| 1 | 763 | 1168 |

cutoff point for our detection model is approximately 0.09 by analyzing the ROC table illustrated in Table 7.11.

Given the optimal cutoff point of 0.09, we create a new Confusion Matrix as illustrated in Table 7.12.

Based on the table above, we can derive the following:

- True positives (TP) is 1168.
- True negatives (TN) is 11,249.

**Table 7.13** Performance metrics

| Performance metrics | Value |
|---|---|
| Accuracy | 0.8196 |
| Misclassification rate | 0.1804 |
| True-positive rate | 0.6049 |
| False-positive rate | 0.1490 |
| True-negative rate | 0.851 |
| False-negative rate | 0.3951 |
| Precision | 0.3722 |
| Prevalence | 0.1275 |

- False positives (FP) is 1970.
- False negatives (FN) is 763.

We now have fewer false negatives, that is, more true positives, which makes sense since we are classifying more of the intrusion data instances. In the meantime, there are, however, many more false positives now. Table 7.13 summarizes a set of metrics that can be derived based on the value of TP, TN, FP, and FN.

## 7.6 Conclusions and Future Work

In this paper, we conducted a detailed analysis of the WUSTL-EHMS-2020 dataset collected on an in-door testbed network including four main components, namely, medical sensors, gateway, network, and control with visualization. The medical sensors are used to collect the patients' vitals. The sensors gather all patients' information and send them through a gateway, during which a man-in-the-middle attack is conducted so attackers can intercept the data being transferred in the middle of the data transfer and then make changes to the patient's information. This can lead to medical professionals giving the wrong treatment to the patients.

The dataset initially includes 40 features in 14,272 normal instances and 2046 attack instances. After conducting a pre-analysis of its descriptive statistics, we downsize it to 19 features in 13,219 normal instances and 1931 attack instances. We then apply a classic logistic regression model to conduct feature reduction where seven significant features are selected from these 19 features. The intrusion classification model is then built based upon the seven significant features. The ROC analysis shows that our detection model achieves an accuracy of 81.96% with an optimal compromise of a true-positive rate of 60.49% and a false-positive rate of 14.9%. Compared to the PCA-based approach applied in the same benchmark dataset, the precision of our prediction model is 37.22%, which is lower than the precision of 88.5% generated by the PCA model. This is simply because the PCA model uses a cutoff value of 400 attack instances; the total number of attack instance identified by the PCA model is only 354 out of 2046, whereas our model can pick up 1168 attacks out of 1931 attack instances, representing a much higher recall rate of 60.49%, compared to the recall rate of 17.3% by PCA model.

The future work includes applying decision trees, clustering, PCA, support vector machines, neural networks, and gradient boosted trees into the same benchmark dataset with seven significant features in a practical intrusion detection system [38] and then conducting comprehensive comparative studies on the detection performance of all these machine learning approaches, where intrusion responses [39] and alert correlation [40] will also be applied to reduce the number of false alarms.

**Acknowledgments** Research supported by New Hampshire- INBRE through an Institutional Development Award (IDeA), P20GM103506, from the National Institute of General Medical Sciences of the NIH, and in partial by Keene State FDG grant.

## References

1. J.J.P.C. Rodrigues, D.B.D.R. Segundo, H.A. Junqueira, M.H. Sabino, R.M.I. Prince, J. Al-Muhtadi, V.H.C. De Albuquerque, Enabling technologies for the internet of health things. IEEE Access **6**, 13129–13141 (2018)
2. M. Papaioannou, M. Karageorgou, G. Mantas, V. Sucasas, I. Essop, J. Rodriguez, D. Lymberopoulos, A survey on security threats and countermeasures in internet of medical things (IoMT). Trans. Emerg. Telecommun. Technol. (2020). https://doi.org/10.1002/ett.4049
3. A. Gatouillat, Y. Badr, B. Massot, E. Sejdić, Internet of medical things: A review of recent contributions dealing with cyber-physical systems in medicine. IEEE Internet Things J. **5**(5), 3810–3822 (2018). https://doi.org/10.1109/JIOT.2018.2849014
4. *Cybersecurity Vulnerabilities Affecting Medtronic Implantable Cardiac Devices, Programmers, and Home Monitors: FDA Safety Communication*. https://www.fda.gov/medical-devices/safety-communications/cybersecurity-vulnerabilities-affecting-medtronic-implantable-cardiac-devices-programmers-and-home. Retrieved on 28 May 2022
5. M. Karageorgou, G. Mantas, I. Essop, J. Rodriguez, D. Lymberopoulos, Cybersecurity attacks on medical IoT devices for smart city healthcare services, in *IoT Technologies in Smart Cities: From Sensors to Big Data, Security and Trust*, (Institution of Engineering and Technology (IET), London, 2020), pp. 171–187
6. R.V. Sampangi, S. Dey, S.R. Urs, S. Sampalli, A security suite for wireless body area networks. arXiv:1202.2171. http://arxiv.org/abs/1202.2171 (2012)
7. W. Lu, Detecting malicious attacks using principal component analysis in medical cyber-physical systems, in *Artificial Intelligence for Cyber-Physical Systems Hardening*, (Springer, 2022)
8. A. Ghorbani, W. Lu, M. Tavallaee, *Network Attacks, Network Intrusion Detection and Prevention: Concepts and Techniques* (Springer, 2009), pp. 1–25. ISBN-10: 0387887709
9. A. Ghorbani, W. Lu, M. Tavallaee, Theoretical foundation of detection, in *Network Intrusion Detection and Prevention: Concepts and Techniques*, (Springer, 2009), pp. 73–114. ISBN-10: 0387887709
10. D. Garant, W. Lu, Mining botnet behaviors on the large-scale web application community, in *Proceedings of 27th IEEE International Conference on Advanced Information Networking and Applications*, Barcelona, Spain, 25–28 March 2013
11. A. Ghorbani, W. Lu, M. Tavallaee, *Detection Approaches, Network Intrusion Detection and Prevention: Concepts and Techniques* (Springer, 2009), pp. 27–53. ISBN-10: 0387887709

12. W. Lu, A. Ghorbani, Bots behaviors vs. human behaviors on large-scale communication networks, in *Proceedings of 11th International Symposium on Recent Advances in Intrusion Detection (RAID 2008)*, ed. by R. Lippmann, E. Kirda, A. Trachtenberg, (MIT, Boston, 2008), pp. 415–416. RAID 2008, LNCS 5230
13. W. Lu, M. Miller, L. Xue, Detecting command and control channel of botnets in cloud, in *Lecture Notes in Computer Science (LNCS)*, vol. 10618, (Springer Nature, 2017), pp. 55–62. ISBN 978-3-319-69154-1
14. W. Lu, A.A. Ghorbani, Botnets detection based on IRC-community, in *IEEE GLOBECOM 2008–2008 IEEE Global Telecommunications Conference*, (2008), pp. 1–5. https://doi.org/10.1109/GLOCOM.2008.ECP.398
15. W. Lu, N. Mercaldo, C. Tellier, Characterizing command and control channel of mongoose bots over TOR, in *3rd International Conference on Wireless, Intelligent and Distributed Environment for Communication. WIDECOM 2020. Lecture Notes on Data Engineering and Communications Technologies*, ed. by I. Woungang, S. Dhurandher, vol. 51, (Springer, Cham, 2020)
16. M. Tavallaee, W. Lu, A. Ghorbani, Online classification of network flows, in *Proceedings of the 7th Annual Conference on Communication Networks and Services Research (CNSR 2009)*, Moncton, New Brunswick, Canada, 11–13 May 2009, pp. 78–85
17. W. Lu, L. Xue, A heuristic-based co-clustering algorithm for the internet traffic classification, in *2014 28th International Conference on Advanced Information Networking and Applications Workshops*, (2014), pp. 49–54. https://doi.org/10.1109/WAINA.2014.16
18. W. Lu, *An unsupervised anomaly detection framework for multiple-connection-based network intrusions* (Ottawa Library and Archives Canada, 2007). ISBN: 9780494147795
19. W. Lu, I. Traore, A new unsupervised anomaly detection framework for detecting network attacks in real-time, in *Cryptology and Network Security*, ed. by Y.G. Desmedt, H. Wang, Y. Mu, Y. Li, (Springer Nature, 2005), pp. 96–109. ISBN 978-3-540-32298-6
20. W. Lu, I. Traore, An unsupervised approach for detecting DDoS attacks based on traffic based metrics, in *Proceedings of IEEE Pacific Rim Conference on Communications, Computers and Signal Processing (PACRIM 2005)*, Victoria, BC, August 2005, pp. 462–465
21. W. Lu, I. Traore, Determining the optimal number of clusters using a new evolutionary algorithm, in *Proceedings of IEEE International Conference on Tools with Artificial Intelligence (ICTAI 2005)*, Hongkong, November 2005, pp. 712–713
22. W. Lu, H.J. Tong, Detecting network anomalies using CUSUM and EM clustering, in *Advances in Computation and Intelligence*, ed. by Z. Li, (Springer Nature, 2009), pp. 297–308. ISBN 978-3-642-04843-2
23. W. Lu, I. Traore, Unsupervised anomaly detection using an evolutionary extension of K-means algorithm. Int. J. Inf. Comput. Secur. **2**(2), 107–139 (2008). Inderscience Publisher
24. W. Lu, I. Traore, A new evolutionary algorithm for determining the optimal number of clusters, in *Proceedings of IEEE International Conference on Computational Intelligence for Modeling, Control and Automation (CIMCA 2005)*, vol. 1, (IEEE Computer Society, 2005), pp. 648–653
25. *WUSTL EHMS 2020 Dataset*. https://www.cse.wustl.edu/~jain/ehms/index.html. Retrieved on 31 May 2022
26. A. Ghorbani, W. Lu, M. Tavallaee, Data collection, in *Network Intrusion Detection and Prevention: Concepts and Techniques*, (Springer, 2009), pp. 55–71. ISBN-10: 0387887709
27. X. Wang, L. Wang, Y. Li, K. Gai, Privacy-aware efficient fine-grained data access control in internet of medical things based fog computing. IEEE Access **6**, 47657–47665 (2018). https://doi.org/10.1109/ACCESS.2018.2856896
28. V.P. Yanambaka, S.P. Mohanty, E. Kougianos, D. Puthal, PMsec: Physical unclonable function-based robust and lightweight authentication in the internet of medical things. IEEE Trans. Consum. Electron. **65**(3), 388–397 (2019). https://doi.org/10.1109/TCE.2019.2926192
29. J. Su, D.V. Vasconcellos, S. Prasad, D. Sgandurra, Y. Feng, K. Sakurai, Lightweight classification of IoT malware based on image recognition, in *2018 IEEE 42nd Annual Computer Software and Applications Conference (COMPSAC)*, (2018), pp. 664–669. https://doi.org/10.1109/COMPSAC.2018.10315

30. A.S. Sangari, J.M.L. Manickam, Public key cryptosystem based security in wireless body area network, in *2014 International Conference on Circuits, Power and Computing Technologies [ICCPCT-2014]*, (2014), pp. 1609–1612. https://doi.org/10.1109/ICCPCT.2014.7054788
31. A.A. Hady, A. Ghubaish, T. Salman, D. Unal, R. Jain, Intrusion detection system for healthcare systems using medical and network data: A comparison study. IEEE Access **8**, 106576–106584 (2020). https://doi.org/10.1109/ACCESS.2020.3000421
32. D. Gupta, M. Gupta, S. Bhatt, A.S. Tosun, Detecting anomalous user behavior in remote patient monitoring, in *2021 IEEE 22nd International Conference on Information Reuse and Integration for Data Science (IRI)*, (2021), pp. 33–40. https://doi.org/10.1109/IRI51335.2021.00011
33. *Network Flow Statistics*, https://unh.box.com/s/m1twmex4ohj1xlosc9s61ivz0f54qmrz. Retrieved on 31 May 2022
34. H.A. Park, An introduction to logistic regression: From basic concepts to interpretation with particular attention to nursing domain. J. Korean Acad. Nurs. **43**(2), 154–164 (2013). https://doi.org/10.4040/jkan.2013.43.2.154. PMID: 23703593
35. A. Ghorbani, W. Lu, M. Tavallaee, Evaluation criteria, in *Network Intrusion Detection and Prevention: Concepts and Techniques*, (Springer, 2009), pp. 161–183. ISBN-10: 0387887709
36. D. Powers, Evaluation: From precision, recall and f-measure to roc., informedness, markedness & correlation. J. Mach. Learn. Technol. **2**(1), 37–63 (2011)
37. K. Nunley, W. Lu, Detecting network intrusions using a confidence-based reward system, in *2018 32nd International Conference on Advanced Information Networking and Applications Workshops (WAINA)*, (2018), pp. 175–180. https://doi.org/10.1109/WAINA.2018.00083
38. A. Ghorbani, W. Lu, M. Tavallaee, Architecture and implementation, in *Network Intrusion Detection and Prevention: Concepts and Techniques*, (Springer, 2009), pp. 115–127. ISBN-10: 0387887709
39. A. Ghorbani, W. Lu, M. Tavallaee, Intrusion response, in *Network Intrusion Detection and Prevention: Concepts and Techniques*, (Springer, 2009), pp. 185–198. ISBN-10: 0387887709
40. A. Ghorbani, W. Lu, M. Tavallaee, Alert management and correlation, in *Network Intrusion Detection and Prevention: Concepts and Techniques*, (Springer, 2009), pp. 129–160. ISBN-10: 0387887709

# Chapter 8
# Weighted Voting Stacking Ensemble Method for Highly Skewed Binary Data Distribution

**Kgaugelo Moses Dolo and Ernest Mnkandla**

## 8.1 Introduction

The need for big data analytics in the banking sector is directly related to the need for complexity of analytical tasks for fraud detection. That is, as the need for big data analytics in the banking sector is becoming more and more important for supporting business operations and financial transactions [1], the need for complex analytical tasks for fraud detection is also becoming crucial for security of business operations and financial transactions.

The NO FREE LUNCH (NFL) theory suggests that no machine learning method can perform well for all measures in every application [2]. Therefore, an evaluation of algorithms is essential for the quality of machine learning methods for revealing predictive data insights to prevent fraudulent transactions of credit cards in South Africa (SA).

Stacking is an ensemble method that uses different machine learning algorithms to generate new dataset that is used in a combiner machine learning method [3]. The combiner machine learning voting method is computed to predict the output. Majority voting is when every model makes a prediction of a transaction, and the class value with more votes is produced as final output. Weighted voting is voting that is undertaken by counting the predictions of better models to predict the output of each transaction. In this study, it is suggested that weighting voting of the classification models requires an optimization of weights to produce good results.

K. M. Dolo (✉) · E. Mnkandla
University of South Africa, Pretoria, South Africa
e-mail: 58527141@mylife.unisa.ac.za; mnkane@unisa.ac.za



I. Woungang, S. K. Dhurandher (eds.), *5th International Conference on Wireless, Intelligent and Distributed Environment for Communication*, Lecture Notes on Data Engineering and Communications Technologies 174,
https://doi.org/10.1007/978-3-031-33242-5_8

Fraud detection refers to the process of detecting fraudulent activities where no prior tendency of fraud exists [4]. To improve the performance of predictive data analysis methods, more advanced performance optimization methods for boosting performance while retaining low variance and low biasness of data are required for the detection of credit card fraudulent activities and thus exposing hidden fraudulent transactions. The performance optimization methods that can adequately optimize parameters of predictive data analysis methods in massive data are becoming more important than ever due to the amount of data being produced in the current era.

The most challenging part of fraud detection involves uncovering fraudulent activities in a complex dataset that has a high number of non-fraudulent transactions and a low number of fraudulent transactions [5]. The dataset complexity in an environment of rapidly growing data introduces inefficiency in fraud detection systems, and the inefficiency acts as an agent that motivates and accelerates the rate of increase of fraudulent transactions in the banking sector. Hence, the banking sector must make a huge investment in complex and efficient analytical systems that detect fraud activities to prevent losses.

### *8.1.1 Stacking Algorithm*

There is no machine learning method that performs well for all measures in every given application according to the NO FREE LUNCH (NFL) theory [2]. It is, therefore, better to use stacking, which is an ensemble method that uses various machine learning algorithms to construct a model that is used to generate a new dataset used in a combiner machine learning algorithm [3]. In the context of binary classification stacking models, different classifier models are trained with the dataset and produce an output that is either 0 or 1. The general idea behind stacked ensembles is to have two or more classifier models that form the base models and a higher level metamodel that uses the predictions of the base models to make a final prediction of the data observation.

The output column (dependent variable) of base classifier models is used as feature of the stacking model. The stacking ensemble technique is used to generate the output value based on the voting method. For each data observation, the stacking model uses either the majority voting method or the weighted voting method for the prediction of the final output value. Majority voting refers to when all base models vote a prediction class of each transaction. On the other hand, weighted voting involves the voting carried out by counting the predictions of better models to predict the output of each transaction. In this research, the weighted voting has been adopted for use in the stacking ensemble model for the classification of the output of each transaction.

#### 8.1.1.1 Weighted Voting Method

It is assumed $\delta_1, \delta_2, \delta_3, \ldots, \delta_n$ is be the weight of transaction, and $\beta_1, \beta_2, \beta_3, \ldots, \beta_n$ is the output values of transactions, where $\beta$ is the output of transaction, $\delta$ is the weight of transaction, and 1, 2, 3, ..., $n$ is the transaction number from 1 to transaction $n$. The term "non-fraudulent" is the opposite of fraudulent. Mathematically, the term "opposite" can be represented by the negation. If the fraudulent transaction is represented by 1, the non-fraudulent transaction can, therefore, be represented by 0.

For weighted voting, the study replaces the output 0 of base models with −1 since direction is important. If the output value of Weighted_Voting is positive then the transaction is fraudulent and if the output value of Weighted_Voting is negative, then the transaction is non-fraudulent. The Weighted_Voting represents the final voting for the output class of the transaction by weight voting the outcome of the base models. The formula for Weight _ Voting is defined as follows:

$$\text{Weighted\_Voting} = \beta_1\delta_1 + \beta_2\delta_2 + \beta_3\delta_3 + \cdots + \beta_n\delta_n \tag{8.1}$$

Furthermore, in this study, it is proposed that weighted voting for the classification models requires an optimization of weights to produce good results. The rest of this paper is organized as follows: In Sect. 8.2, the research questions addressed by this work and their objectives are presented. In Sect. 8.3, the research presents the research design. Section 8.4 presents the simulation results, and Sect. 8.5 concludes this paper.

## 8.2 Research Questions and Objectives

### *8.2.1 Research Questions*

Since the South African banking sector is migrating from data analytics to data insights [6], the study, therefore, sets out to explore the following research question:

(a) Does manual stacking of feature selection methods, which are diverse on the Principal Component Analysis (PCA) transformed features to hide the actual feature names, select enough good features to solve the problem than just select a feature engineering technique?
(b) Does the stacking of supervised base models have an impact on predictive accuracy given the undersampling technique to remove the duplicate data observations on the majority class and the Safe-Level-SMOTE oversampling method?

This study, therefore, sought to answer these questions by carrying out the objectives presented in the next section.

### *8.2.2 Research Objectives*

The main research objective is to obtain optimum data analytics algorithms that perform data insights to reveal fraudulent data patterns in order to prevent future fraudulent transactions. The sub-objectives are as follows:

(a) To determine whether manual stacking of diverse feature selection methods on the PCA transformed features to solve the problem than feature selection method. This objective will be implemented by running univariate selection algorithm, recursive feature elimination algorithm, and feature importance algorithm on the entire dataset.
(b) To investigate how the stacking of supervised machine learning algorithms have an impact on predictive accuracy, considering the down-sampling technique, which eliminates duplicate data samples on the class of majority data observations, and the SMOTE oversampling method. This objective will be implemented by running both the differential evolution optimization and the stacking algorithms on the classifier models.

The next section discusses the research design followed to achieve the research objectives of this study.

## 8.3 Research Design

The weighted stacking ensemble base models' performance will be compared with optimized weighted stacking ensemble base models that represent both classes so well. The credit card fraud dataset was collected, preprocessed, and normalized. To solve the issue of class imbalance, the modified Safe-Level-SMOTE was used to oversample the data observation of the minority class of the dataset. The oversampling was done together with the down-sampling method that removes duplicates and randomly chooses nonduplicate data samples from the majority class in order to control the best fit line to an optimum place that represents fraudulent and non-fraudulent transactions equally [7]. The feature selection was performed using the manual stacking ensemble method of Univariate Selection, Recursive Feature Elimination, and Feature Importance. Thus, the weighted stacking ensemble base models' performance and the optimized weighted stacking ensemble base models' performance were tested by running Artificial Neural Network, Support Vector Machine, Naïve Bayesian, and k-Nearest Neighbors algorithms.

### 8.3.1 *Stacking Ensemble of Base Models*

The voting class named Voting, which is defined below, allows base models to vote the final output of each transaction of credit card data. The final output value of each transaction can be classified as either 1 to denote a fraudulent transaction or 0 to denote a non-fraudulent transaction. The voting method that has been adopted for this study is based on the principle of majority rule, which means that the class with the majority of votes in respect of base models is the final classification of the transaction. The following pseudocode shows the flow of how the Voting Method of the stacking ensemble method was defined.

**Stacking Ensemble Method**

```
Input:
Output: Predicted Output
Begin
 1. Function to return voted class (Modc_Fun)
    Input: List
    Predict, Count0, Count1 = 0, 0, 0
    For each value in the List
            If value = 0
                    Count0 += 1
            Esif value = 1
                    Count += 1
    If Count0 > Count1
            Predict = 0
    Else
            Predict = 1
    Return Predict
 2. Function to return list of predicted values
    Input: List of Base model predicted values
    (Model_pred_A)
    Predictions = []
    For each Row in Model_pred_A
          Mode =  Mode_fun(Row)
          Append Mode into Predictions
    Return Predictions
 3. Function to get an Accuracy
    Input: TestDataY, predictions
    Correct = 0
    For y = 0 to length of TestDataY
          If TestDataY[y] = predictions[y]
              Correct += 1
    Return (Correct / length of TestDataY) * 100
 4. Function to return unmatched Indexes
    Input: TestDataY, Predictions
    Incorrect = []
    For ind = 0 to length of TestDataY
```

(continued)

```
                If TestDataY[ind] is not equal to
                Predictions[ind]
                      Append ind into Incorrect
        Return Incorrect
   End of the Class
```

To enhance the performance of the voting method, the weighted voting technique is required. In this study, the weighted voting method was performed using the predicted class probabilities of output values generated by the base models. Probability is known to be the number between 0 and 1 indicating the likelihood of an event occurring, and class value is the number that represents the class in which the probability value is closer to. This means that, in this study, the predicted class probability of an output class was used for weighted voting method.

### 8.3.2 Probability Data Dictionary

The use of the Stacking Ensemble method in combination with weighted voting of the base data models involves creating a new data dictionary of base models that is composed of class probability of the predicted value of each credit card transaction. The data dictionary name is Model_Probability and is defined as follows:

$$ANN_{pred} = nn.\,predict(X_{test})$$
$$ANN_{pred\_Avg} = nn.\,predict\_proba(X_{test})$$
$$AnnProbability = nn.\,probability(Ann_{pred}, Ann_{pred\_Avg})$$
$$SVM_{pred\_Avg} = nnSVM.\,predict\_proba(X_{test}, y_{test})$$
$$SVM_{pred} = nnSVM.\,predict(X_{test}, y_{test})$$
$$SVMProbability = nnSVM.\,probability(SVM_{pred}, SVM_{pred\_avg})$$
$$KNNProbability = nnKNN.\,probability(np.\,array(X_{test}))$$
$$clf_{object} = nnDT.\,train\_using\_entropy(Xs, Ys)$$
$$DT_{pred\_proba} = nnDT.\,pred\_proba(X_{test}, clf_{object})$$
$$DT_{pred} = nn.\,prediction(X_{test}, clf_{object})$$
$$DT_{pred} = [1\ if\ i == 1.0\ else\ 0\ for\ i\ in\ DT\_pred]$$
$$DTProbability = nnDT.\,probability(DT_{pred}, DT_{pred\_avg})$$
$$NBProbability = nnNB.\,probability(X_{test})$$
$$Model_{probability} = \{ 'ANN' : AnnProbability,\ 'SVM' : SVMProbability,\ 'KNN' : KNNProbability,\ 'DT' : DTProbability,\ 'NB' : NBProbability \}$$

The nn, nnSVM, nnKNN, nnDT, and nnNB are the class objects of the base models defined in this study. The Model_Probability is the new data dictionary that is composed of predicted class probability and base model names. In the data analysis section below, the study compares the weighted stacking ensemble base models' performance and the optimized weighted stacking ensemble base models' performance.

## 8.4 Simulation Results

### 8.4.1 In Accuracy of Base Models

The base models in this study are Artificial Neural Network, Support Vector Machine, Naïve Bayesian, and k-Nearest Neighbors algorithms. The Python code below prints the name and the overall accuracy score of the base models. nn, nnSVM, nnKNN, nnDT, and nnNB are the class objects. The score and getAccuracy are the class properties of base models. The $X_{test}$ and $y_{test}$ are the respective input and output values of the test data. The accuracy of the base model was achieved by running the following code:

$print('Artificial\ Neural\ Network:' + str(nn.\ score(X_{test}, y_{test})))$
$print('Support\ Vector\ Machine:' + str(nnSVM.\_score(X_{test}, y_{test})))$
$print('K - Nearest\ Neighbor:'$
$\quad + str(nnKnn.\_getAccuracy(list(X_{test}, predProb)))$
$print('Decision\ Tree:' + str(nnDT.\_getAccuracy(list(y_{test}), predProb)))$
$print('Naive\ Bayesian:' + str(nnNB.\_getAccuracy(list(y_{test}), Predicted_y)))$

Output:
Artificial Neural Network: 0.9518275341455708
Support Vector Machine: 0.7108598714932762
K-Nearest Neighbor: 90.83482555621877
Decision Tree: 99.17488852217268
Naive Bayesian: 97.06002832297555

When compared with other base models, the Decision Tree Method performed extremely well and achieved a higher accuracy rate. The Support Vector Machine method achieved an accuracy that is far lower than that of other base models, whereas the overall performance of all the base models are shown to classify majority of the data observations correctly. The statistics show which class of the distribution is well classified is not shown.

**Table 8.1** Confusion matrix

| Model | TP | TN | FP | FN |
|---|---|---|---|---|
| ANN | 5 | 81,130 | 4176 | 132 |
| SVM | 22 | 61,753 | 23,553 | 115 |
| KNN | 7 | 77,359 | 7947 | 130 |
| DT | 0 | 84,492 | 814 | 137 |
| NB | 4 | 82,699 | 2607 | 133 |

**Table 8.2** Confusion matrix statistics

| Model | TP | TN | FP | FN |
|---|---|---|---|---|
| ANN | 0.036496 | 0.951047 | 30.481752 | 0.001547 |
| SVM | 0.160584 | 0.723900 | 171.919708 | 0.001348 |
| KNN | 0.051095 | 0.906841 | 58.007299 | 0.001524 |
| DT | 0.000000 | 0.990458 | 5.941606 | 0.001606 |
| NB | 0.029197 | 0.969439 | 19.029197 | 0.001559 |

### 8.4.2 *Confusion Matrix*

The confusion matrix is a classification model evaluation method that shows the total number of true negative, true positive, false negative, and false positive. The number of true-positive (TP) transactions is the number of output transactions that the base classifier has predicted as being fraudulent when they were indeed fraudulent. Similarly, the count of true-negative (TN) transactions is the count of output transactions that the base classifier has predicted as being non-fraudulent when they were indeed non-fraudulent. The count of false-positive (FP) transactions is the count of output transactions that the base classifier has predicted as being fraudulent when they were in fact non-fraudulent, and the count of false-negative (FN) transactions is the count of output transactions that the base classifier has predicted as being non-fraudulent when they were in fact fraudulent. The total number of illegal and legal transactions for testing dataset was 137 and 85,306.

Table 8.1 lists the negative (legal) and positive (illegal) transactions to clearly show the base models are extremely biased. To give the above confusion matrix table meaning, the output of confusion matrix can be illustrated in statistical form as shown in Table 8.2.

The negative (FN) transactions listed in Table 8.2 shows that the Decision Tree method had a higher percentage of accuracy compared to other base models. However, the Decision Tree method was extremely biased compared to other base models; the Support Vector Machine method had a lower percentage of accuracy when compared with other base models. On the other hand, the true positive (TP) of Support Vector Machine method is greater than the true positive value of other base models.

Therefore, it can be concluded that a higher percentage of accuracy of the base model that is extremely biased is meaningless compared to an average percentage of accuracy of the base model where all output class values are well represented.

In this study, the true accuracy of a base model is the accuracy in which all output classes are well represented.

The stacking ensemble base models were optimized to achieve equal representation of output classes while ensuring that the high overall accuracy of base models is maintained. The optimization methods of stacking ensemble base models are discussed in the next section.

### 8.4.3 *Optimized Stacking Ensemble Base Models*

To obtain the true accuracy of base models and thus achieve a classification problem of equal representation of class values, the hyperparameters of the base models were modified. Furthermore, a classification problem of equal representation of class values was optimized to maintain the higher overall accuracy of base models. The hyperparameter tuning of the base models was done using a grid search method. Once all the base models were optimized, the class object were used in combination with the class properties to rerun the accuracy code and print the true accuracies of the base models. The true accuracies of the base models are detailed in the following subsection.

#### 8.4.3.1 True Accuracy of Base Models

The python code shown below was used to prints the name and the overall true accuracy score of the base models. The class object and the class properties were used to run the accuracy code to print the true accuracies of the base models. Whereas the $X_{test}$ refers to the input values of the test data, the $y_{test}$ refers to the output values of the test data. The true accuracy of the base model was achieved by running the following code:

$print('Artificial\ Neural\ Network:' + str(getAccuracy(Ann_{pred}, np.\ array(y_{test}))))$
$print('Support\ Vector\ Machine:' + str(getAccuracy(SVM_{pred}, np.\ array(y_{test}))))$
$print('K - Nearest\ Neighbor:' + str(getAccuracy(knn_{pred}, np.\ array(y_{test}))))$
$print('Decision\ Tree:' + str(getAccuracy(DT_{pred1}, np.\ array(y_{test}))))$
$print('Naive\ Bayesian:' + str(nnNB.\_getAccuracy(list(y_{test}), NB_{pred})))$
Output:
Artificial Neural Network: 95.19562749435296
Support Vector Machine: 72.8637805320506
K-Nearest Neighbor: 92.20650023992604
Decision Tree: 95.18158304366654
Naive Bayesian: 96.94065049214096

The Naïve Bayesian method performed extremely well; relative to other base models, a higher accuracy rate was achieved. The accuracy of the SVM method was

**Table 8.3** True confusion matrix

| Model | TP | TN | FP | FN |
|---|---|---|---|---|
| ANN | 98 | 81,240 | 4066 | 39 |
| SVM | 136 | 62,121 | 23,185 | 1 |
| KNN | 126 | 78,658 | 6648 | 11 |
| DT | 125 | 84,711 | 595 | 12 |
| NB | 124 | 82,705 | 2601 | 13 |

**Table 8.4** True confusion matrix statistics

| Model | TP | TN | FP | FN |
|---|---|---|---|---|
| ANN | 0.715328 | 0.952336 | 29.678832 | 0.000457 |
| SVM | 0.992701 | 0.728214 | 169.233577 | 0.000012 |
| KNN | 0.919708 | 0.922069 | 48.525547 | 0.000129 |
| DT | 0.912409 | 0.993025 | 4.343066 | 0.000141 |
| NB | 0.905109 | 0.969510 | 18.985401 | 0.000152 |

far lower than those of other base models. To check how well the class values of base models are represented, the confusion matrix is required. The true confusion matrix is discussed in the subsection that follows.

#### 8.4.3.2 True Confusion Matrix

This shows the number of true-negative, true-positive, false-negative, and false-positive transactions, which are defined in Sect. 8.4.2. The confusion matrix function was used to generate the following confusion matrix shown in table defined in Sect. 8.4.3 of this study. For the testing of the dataset, the total count of illegal transactions and the total count of legal transactions of 137 and 85,306 were used, respectively (Table 8.3).

To give the above confusion matrix more meaning, the confusion matrix statistics is required. To display the confusion matrix statistics, the confusion matrix statistics function is defined, and the output of the confusion matrix statistics is shown in Table 8.4.

The confusion matrix statistics in Table 8.4 shows that the true positive and negative transactions are well represented. This means that the fraudulent data and non-fraudulent data instances are well classified. Since the true accuracy of a base model is the accuracy in which all output classes are well represented, it can be concluded that the accuracies in Sect. 8.4.3 are true accuracies of base models. The true accuracies are required to model a stacking ensemble method. Stacking ensemble method is a classification method that uses more than one classification model for decision-making, and it is discussed in the next section.

### 8.4.4 Stacking Ensemble Method

The Stacking Ensemble Method is the final voting of the class values generated by the base modes. The voting class named Voting allows base models to vote the final output of each transaction of credit card data. To run the Voting class code, the class Voting() was used as an object. The class object contains the class definition called class properties. The Model_predictions data dictionary that contains predicted class values by the base models was used for voting of the final class values. The following class object was used to return a list of predicted output values:

$$Final_{pred} = Voting\left(\right).VoteFunc\left(np.array\left(pd.DataFrame\left(Model_{predictions}\right)\right)\right)$$

Vote_Func is the class property of Voting class, which returns the list of final prediction of class values that belongs to credit card transactions. The list of predicted class values was required to compute the accuracy of stacking ensemble method. The accuracy of the stacking ensemble method that represents the class values of the dataset well while maintaining or enhancing the performance of the classification model is essential to gain a confidence in the classifier model. The accuracy of the stacking ensemble method is discussed in the next subsection.

#### 8.4.4.1 Stacking Ensemble Method Accuracy

The stacking ensemble method accuracy was the accuracy that represents the class values of the dataset well while maintaining or enhancing the performance of the classifier model. The voting class property of accuracy was getAccuracy which takes two parameters, namely, the test output values of the dataset named $y_{test}$, and the predicted output of credit card transactions named $Final_{pred}$. To display the accuracy of the stacking ensemble method, the following python code was executed:

$$Voting\left(\right).getAccuracy\left(np.array\left(y_{test}\right), Final_{pred}\right)$$

The output of 99.39842936 suggested that the stacking ensemble method was performing well. However, to determine how well the stacking ensemble method was performing, the confusion matrix was required. The confusion matrix of stacking ensemble method is discussed in the subsection that follows.

#### 8.4.4.2 Confusion Matrix of Stacking Ensemble

This shows the number of true-negative, true-positive, false-negative, and false-positive transactions. The confusion matrix was discussed in more detail in Sect. 8.4.2. The output of confusion matrix is given in Table 8.5:

**Table 8.5** Confusion matrix of stacking ensemble

| Model | TP | TN | FP | FN |
|---|---|---|---|---|
| Stacking Ensemble | 92 | 84,837 | 367 | 147 |

**Table 8.6** Confusion matrix statistics of stacking ensemble

| Model | TP | TN | FP | FN |
|---|---|---|---|---|
| Stacking Ensemble | 0.6715329 | 0.994502 | 2.678832 | 0.0017232 |

The total number of illegal and legal transactions is 137 and 85,306, respectively. The output of stacking ensemble method confusion matrix shows that the both base models and stacking ensemble method are performing well. However, to enhance the understanding of the confusion matrix, the statistics of the confusion matrix was required. The statistics of the stacking ensemble method confusion matrix is discussed in the following subsection.

#### 8.4.4.3 Confusion Matrix Statistics of Stacking Ensemble

The statistics function of the stacking ensemble method confusion matrix was designed to show how well the class values are represented (in a form of percentages) by the prediction model. The output confusion matrix statistics of stacking ensemble is displayed in Table 8.6.

The total number of illegal transactions was found to be 137, and the total number of legal transactions was 85,306. The statistics of the stacking ensemble method confusion matrix shows how well the class values are represented by the prediction model. However, a good classification model minimizes the count of false negatives and positives, even when the count of true negatives and positives is good in relation to the total count of transactions.

Therefore, this research study sought to optimize the stacking ensemble method to decrease the number of misclassified transaction while enhancing the performance of the stacking ensemble method.

## 8.5 Results

(a) First Subresearch Question

"Does manual stacking of feature selection methods, which are diverse on the PCA transformed features to hide the actual feature names, select enough good features to solve the problem than just selecting a feature selection algorithm?"

The manual stacking of feature selection methods that are diverse on the PCA transformed features performed differently for each of the features. However, similarities were noted for features that were not adding or adding less value to the classification problem. The time feature appeared to be adding less value to the classification problem when computing univariate selection algorithm, recursive feature elimination algorithm, and feature importance algorithm. For this reason, the time feature was removed from the dataset.

(b) Fourth Subresearch Question

"Does the stacking of the supervised machine learning algorithms have an impact on predictive accuracy given the undersampling technique to remove the duplicate data observations on the majority class, and the Safe-Level-SMOTE oversampling method?"

The stacking of the supervised machine learning algorithms has proven to positively impact the predictive accuracy given the undersampling technique to remove the duplicate data observations on the majority class and the Safe-Level-SMOTE oversampling method. Like the base models (machine learning methods), the stacking method was found to be extremely biased.

However, the stacking method had fewer false negatives and false positives compared to the base models. This means that the stacking ensemble method performed slightly better than the base models. Therefore, the claim about the majority voting stacking ensemble method not being a good stacking ensemble method for the credit card dataset as compared to the weighted voting stacking ensemble method was correct.

## 8.6 Conclusion

In conclusion, it can be said that the Artificial Neural Network, Support Vector Machine, Naïve Bayesian, and k-Nearest Neighbors algorithms are powerful binary classification models. However, in a highly skewed binary classification problem, the binary classification models were observed to misclassify an adequate number of minority class observations (illegal transactions). Applying an optimized stacking ensemble model to weight vote the credit card transactions was observed to improve the performance by misclassifying very few illegal transactions while classifying a huge percentage of legal transactions.

The study, therefore, concludes that the proposed weighted stacking ensemble model, which was evaluated using the confusion matrix, was found to perform better than the base models in terms of accuracy. Furthermore, the model was found to cover almost the entire search space of the credit card transactions. However, it is noteworthy that the proposed model has its own limitations. Specifically, the framework for guiding the study in terms of the selection for stacking ensemble method of the base models covering the entire search space of the credit card data observations was not addressed.

**Acknowledgments** I would like thank Prof Ernest Mnkandla for providing insight, expertise, and comments that greatly assisted the research. I would also like thank my family: Joyce Dolo, Mahlatse Dolo, Kamogelo Dolo, and Tshegofatso Atlegang Kgoele for their support.

## References

1. M. Altayar, N. Almoqren, The motivations for big data mining technologies adoption in Saudi banks, in *4th Saudi International Conference on Information Technology (Big Data Analysis) (KACSTIT)*, (IEEE, 2016), pp. 1–8
2. W. Macready, D. Wolpert, No free lunch theorems for optimization. IEEE Trans. Evol. Comput. **1**, 67–82 (1996)
3. V. Smolyakov, *Ensemble Learning to Improve Machine Learning Results*. [Online]. Available at: https://blog.statsbot.co/ensemble-learning-d1dcd548e936 (2017)
4. S. John et al., Realtime fraud detection in the banking sector using data mining techniques/algorithm, in *International Conference on Computational Science and Computational Intelligence*, (IEEE, 2016), pp. 1186–1191
5. M. Pushpa, N. Malini, Analysis on credit card fraud identification techniques based on KNN and outlier detection, in *3rd International Conference on Advances in Electrical, Electronics, Information, Communication and Bio-Informatics (AEEEICB17)*, (IEEE, 2017), pp. 255–258
6. TCI, *Driving profitable growth through improved data insights and segmentation practices for the future customer and digital bank* (Indaba Hotel–Sandton, Trade Conferences International, 2015), pp. 1–7
7. E.M. Kgaugelo Moses Dolo, Modifying the SMOTE and safe-level SMOTE oversampling method to improve performance, in *4th International Conference on Wireless, Intelligent and Distributed Environment for Communication*, (Springer, 2022), pp. 47–59

# Chapter 9
# Hyperparameter Tuning for an Enhanced Self-Attention-Based Actor-Critical DDPG Framework

**Ehsan Ur Rahman Mohammed, Surajsinh Prakashchandra Parmar, Rayhaan Pirani, and Kriti Kapoor**

## 9.1 Introduction

Reinforcement learning is a highly critical learning paradigm in artificial intelligence. It suits all those scenarios where we do not find sufficient training data that makes the requirement of training necessary based on experience. Despite huge successes by researchers in this area, there is still a long way to go to deal with problems such as sub-optimal performance, slow training, instability in learning, and many more. The use of applications and tools such as self-driving cars requires more research on reinforcement learning. In the coming days, with learning various human-level tasks becoming common and insufficient data to train models, reinforcement learning will become the go-to solution of the hour. The relative newness of reinforcement learning (RL), despite its challenges while applying to real-world problems as mentioned by Dulac-Arnold et al. [1], and the scope of improving DDPG have motivated us to perform this study.

We shall focus on one of the recently famed approaches to RL, called the Deep Deterministic Policy Gradient (DDPG) algorithm. Briefly speaking, a model-free off-policy approach for learning continuous actions is Deep Deterministic Policy Gradient (DDPG). A model-free approach means it does not have a data model to use for making decisions. Instead, it can have a policy or set of rules or heuristics to make decisions. It combines Deterministic Policy Gradient (DPG) and Deterministic Quality Network (DQN) principles. DPG helps in operating/learning

E. U. R. Mohammed (✉) · S. P. Parmar · R. Pirani · K. Kapoor
School of Computer Science, University of Windsor, Windsor, ON, Canada
e-mail: rahman6a@uwindsor.ca; parmar64@uwindsor.ca; piranir@uwindsor.ca; kapoor13@uwindsor.ca



I. Woungang, S. K. Dhurandher (eds.), *5th International Conference on Wireless, Intelligent and Distributed Environment for Communication*, Lecture Notes on Data Engineering and Communications Technologies 174,
https://doi.org/10.1007/978-3-031-33242-5_9

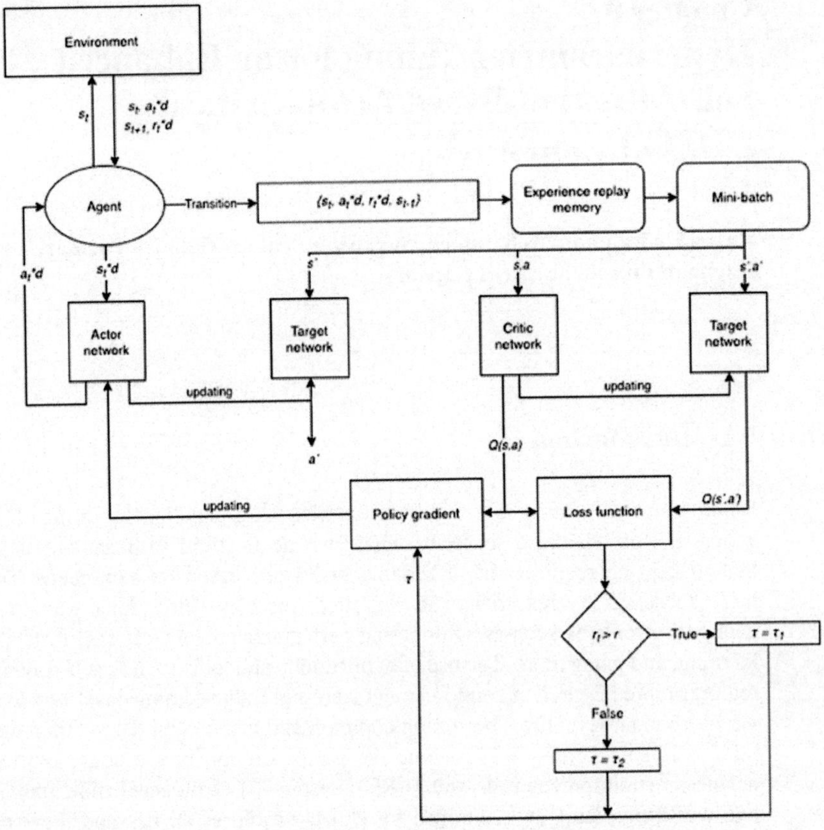


**Fig. 9.1** Overview of our proposed work

over continuous action spaces, DQN gives the experience replay mechanism and slow-learning target networks.

Figure 9.1 can very well summarize and represent the central concept of DDPG. It comprises two network architectures in the actor-critic framework as the policy network piece. One is called the actor network, which outputs a given action based on the current state, and the next is the critic network that predicts whether the given action is good or bad given the action and state pair. We also have another target network piece that computes the Q-value function and thereby adds stability during training. It is only updated with the policy network's weights now and then.

## 9.2 Related Works

Wu and Li [2] present a policy learning mechanism called aggregated multi-DDPG (AMDDPG) for a self-driving policy using the TORCS software. The paper's authors deploy multiple DDPGs to learn multiple subpolicies, which are then aggregated at a later stage to give the aggregated policy. A centralized experience reply buffer helps break any correlation between the samples and helps an agent utilize the experiences from the multiple subpolicies. An ablation study is also conducted to measure the generalization, effects of aggregation, and effects of the number of subpolicies. This method converges approximately 57% faster than the original DDPG algorithm and gives a better generalization performance. However, the presence of multiple conflicting policies and the need to learn more policies makes the model computationally expensive.

DDPG suffers from problems like long training time. The work presented in Xiang et al. [3] overcomes some of the problems with DDPG by using a two-way search and a novel reward mechanism for the task of obstacle avoidance. An actor-critical framework-based DDPG algorithm is constructed using a dual neural network. The dual network has the same structure with different estimated and target network parameters. Even though the authors of Xiang et al. deploy an experience replay buffer to enhance the utilization rate of samples, the ability to have a two-way search in real-world scenarios is limited. A double-network prioritized experience replay mechanism (DNPER-DDPG) is proposed by Kang et al. [4]. The value function is approximated using two neural networks. The updated value of the actor policy network is chosen as the minimum of the action value functions generated by the two networks, reducing the incidence of local optimum policy. The Q-values obtained by the two networks combined with the immediate reward obtained by the environment are utilized as a criterion for prioritizing samples from the experience replay mechanism to increase the algorithm's convergence speed. Our work tends to better this approach by incorporating some novelties as will be mentioned in the Sect. 3.

Jesus et al. [5] discuss using Deep Deterministic Policy Gradient (DDPG) networks to navigate mobile robots. This network considers inputs like the robot's position, orientation, and angular and linear velocities. It provides the angular and linear velocities that can be used as output control signals for the robot. The authors performed experiments on two natural environments using physical robots. The paper concluded that using DDPG networks is highly effective for the continuous control of robots. They also noted that this approach requires a significant amount of time to train the model to navigate, which can be aided using streamlined training approaches such as ours. Though we have not experimented extensively on training time, our initial results are very promising in reducing the training time.

Liu et al. [6] studied an implementation of a bipedal walking controller in the GAZEBO simulation model based on an average male designed to build a walking system to support people with spinal cord injuries to walk as fast as 1 m/s for extended periods. This simulation predicted the ideal foot placement needed to maintain stability using a DDPG network. Even after applying some impulse on the torso and rear, the controller showed stability with backpacks and weights added to the legs. It demonstrated robustness by offering stability for different heights and weights. A parallel DDPG approach is presented in Tao et al. [7] for biped robot gait control. Multiple Actor-Critic (AC) structures realize the DDPG algorithm's parallelization. An optimized experience replay mechanism is used to enhance the quality of training. The actions and the state space were narrowed down using knowledge and experience. The parallel DDPG algorithm, when applied to the biped robot, demonstrated a more considerable stride length and more stable hip change, resulting in the robot walking faster and with more stability.

Studies like Hauser [8] speak about establishing sufficient consensus on how essential hyperparameters and their tuning is regarded in RL. Solving hyperparameter tuning by framing it as a multiarmed bandit problem makes the problem shift into RL space, gives a different perspective, and opens other opportunities to explore it. Wu and Li [2] also explore the idea of reusing samples collected by "one configuration to estimate the value of a different configuration to improve sample efficiency." The choice of hyperparameters is a critical step in any machine learning experiment. Nesma et al. [9] use a swarm-based optimization technique for hyperparameter tuning, which they call "Whale Optimization Algorithm (WOA)." WOA is explicitly deployed to fine-tune the hyperparameters of the (DDPG) algorithm to learn an optimal control strategy for self-driving vehicles' control. DDPG can handle complex environments, which contain continuous spaces for actions. Mishra and Prakash [10] put in place DDPG along with PPO as a means to combine off-policy and on-policy algorithms to address the task of making bipedal walkers learn efficiently. They also propose a model-free control framework to tackle the issue of finding the best-performing hyperparameters.

In the study by Kumar et al. [11], the authors use GAZEBO, which is a realistic robotics simulator; a planar bipedal walking robot (BWR) was developed and simulated. The BWR used DDPG and an adequately structured reward function and successfully made walking of the bipedal walker quicker, resembling that of a human. In the study performed by Ahmadi et al. [12], CACLA (Continuous Actor-Critic Learning Automaton) and SPG (Sampled Policy Gradient), which are two RL algorithms designed to work with continuous input and action spaces, are presented and tested with the OpenAI Gym environment. The OpenAI Gym platform is used to test these two algorithms. The BipedalWalker-v2 environment of OpenAI Gym was selected to conduct the training and testing of both algorithms. CACLA showed improved performance, and the implemented SPG showed difficulties.

In the research by Grbic and Risi [13], current RL techniques only allow the agent to adjust during a specific training period, not during execution. The 2D bipedal walker domain employs a method that combines neuro-evolution to select a deep neural network's beginning weights with a variant of deep reinforcement learning that runs continuously during execution time. The results suggest that the combination strategy outperforms evolution or reinforcement learning alone in generalization. Combining reinforcement learning with evolution opens up intriguing new research avenues for constantly learning agents that can benefit from appropriate priors chosen by an evolutionary process.

In Ha [14], the prospect of concurrently learning a better-suited version of the agent's design and the policy are investigated in great detail. The theory is achieved with the proposal of a modification to the popular OpenAI Gym framework. The authors have parameterized portions of an environment and allowed the agent to learn to change these environment parameters in tandem with its policy. This makes RL faster and policy learning easier. Joint policy and structure learning might lead to the discovery of design principles that could be valuable in assisted-design applications in future application areas of RL as described in Hou et al. [15].

## 9.3 Proposed Work

We have used the pendulum environment for experimenting with our proposed work. The environment was accessed from the OpenAI's Gym library. The implementation was done in Python, with other deep learning packages like PyTorch for creation, training, and testing of the actor-critical DDPG framework. Pendulum is a single-agent RL environment with the action space comprising of torque values and the reward being the negative of the cosine of the angle between the pendulum and the horizontal axis.

The problem we are tackling revolves around creating a more robust actor-critical network architecture and tuning the training hyperparameters for an efficiently trained policy. Experience replay also enhances the training by helping the network converge faster through the recorded training episodes.

DDPG, which stands for Deep Deterministic Policy Gradient, is a specialized algorithm used to learn policies for RL agents. DDPG is a model-free off-policy algorithm for learning continuous actions as cited by Kuznetsov et al. [16]. It combines the two ideas from DPG (Deterministic Policy Gradient) and DQN (Deep Q-Network). It uses Experience Replay and slow-learning target networks from DQN, and it is based on DPG, which can operate over continuous action spaces. The policy is deterministic since it directly outputs the action. In order to promote exploration, some Gaussian noise is added to the action determined by the policy. To calculate the Q-value of a state, the actor output is fed into the Q-network.

Actor-critical network comprises of two networks; one is called as the actor network that is meant to propose an action for the agent in a given state. The other network is crucial in learning the optimal policy as it predicts whether the given action proposed by the actor network is good (positive or higher reward) or bad (negative or lower reward) given the state of the environment. This actor-critical framework is one of the things that differentiates DDPG from the original DQN algorithm. Using actor and critic networks helps in adding stability to training, as we are learning from estimated target values, and hence, if we update the target network slowly, we can keep our estimated targets stable.

Experience replay is the task of storing a list of tuples with state, action, reward value, and next state. This list if used to learn the policy not only from recent experience but also from a sample of all of our experience accumulated so far. This helps in efficient use of previous experience, which is considered to be the key when gaining real-world experience is costly, as mentioned in the literature. Another great benefit is faster convergence. Parameters that define the architecture of a learning model or shape its training mechanism are called as hyperparameters. The process of automated searching for the ideal set of hyperparameters for a given model architecture, a given dataset, and a given task is called hyperparameter tuning. The hyperparameters that were selected for our training are learning rate, weight initialization, and weight decay.

We have also tried embedding different types of layers into the actor-critical network such as one-dimensional convolution, dropout, batch normalization layers, among others. We have applied a small amount of discount factor to the rewards stored in the experience replay buffer, and while using the experience replay mechanism, we most of the times selected from the subset of experiences having a higher reward and less often selected other experiences. We tested a self-attention-based architecture for both the actor and the critic networks along with a self-attention architecture for the target network, tried the various combinations such as having self-attention-based architecture for target and not for the policy network, vice versa, and had fully connected network components in both networks. The results suggest that xyz combination works better than the rest, and this can be attributed to the advantages of the abc architecture. The way the weights of the policy network are updated based on the updates of target network in the previous literature does not take into account the loss; here based on the value of loss, we change the tau value, higher tau if the loss is higher and lower tau if the loss is lower.

Algorithm illustrated in Fig. 9.2 and flowchart as illustrated in Fig. 9.1 describe our proposed version of the DDPG algorithm used in our implementation. Cross-validation for hyperparameter tuning is performed to obtain the consolidated weight metric. This metric can be compared to find the best parameter values, as presented in the Figs. 9.3 and 9.4.

## 9.4 Experiments and Results

### *9.4.1 Experimental Setup*

The created pendulum environment with the goal of defying gravity by applying torque has an action space of one-dimensional value in the range of [−2, 2], which represents torque values in clockwise and counterclockwise directions. We decided to experiment with hyperparameters in the Agent and Critic networks, which leads us to the best scores. The hyperparameters selected for a grid search-based

**Algorithm 1** Proposed Work

**procedure** PROPOSED WORK
    Randomly initialize critic self-attention network $Q(s, a|\theta^Q)$ and actor self-attention network $\mu(s|\theta^\mu)$ with weights $\theta^Q$ and $\theta^\mu$.
    Initialize target self-attention network $Q'$ and $\mu'$ with weights $\theta^{Q'} \leftarrow \theta^Q$, $\theta^{\mu'} \leftarrow \theta^\mu$
    Initialize replay buffer $R$
    Initialize discount factor $d$, $\tau_1$, $\tau_2$
    Initialize reward threshold $r$
    **for** episode = 1, M **do**
        Initialize a random process $N$ for action exploration
        Receive initial observation state $s_1$
        Select action $a_t = \mu(s_t|\theta^\mu) + N_t$ according to the current policy and exploration noise (OU noise)
        Execute action $a_t$ and observe reward $r_t$ and observe new state $s_{t+1}$
        **if** $r_t > r$ **then**
            $\tau \leftarrow \tau_1$
        **else**
            $\tau \leftarrow \tau_2$
        Store transition $(s_t, a_t * d, r_t * d, s_{t+1})$ in $R$
        Sample a random minibatch of $N$ transitions $(s_i, a_i, r_i, s_{i+1})$ from $R$
        Set $y_i = r_i + \gamma Q'(s_{i+1}, \mu'(s_{i+1}|\theta^{\mu'})|\theta^{Q'})$
        Update critic by minimizing the loss $L = \frac{1}{N}\sum_i (y_i - Q(s_i, a_i|\theta^Q))^2$
        Update the actor policy using the sampled policy gradient

$$\nabla_{\theta^\mu} J \approx \sum_i \frac{1}{N} \nabla_a Q(s, a|\theta^Q)|_{s=s_i, a=\mu(s_i)} \nabla_{\theta^\mu} \mu(s|\theta^\mu)|_{s_i}$$

        Update the target networks:

$$\theta^{Q'} \leftarrow \tau\theta^Q + (1-\tau)\theta^{Q'}$$

$$\theta^{\mu'} \leftarrow \tau\theta^\mu + (1-\tau)\theta^{\mu'}$$

    **return** consolidated weights

**Fig. 9.2** Algorithm of our proposed work

hyperparameter tuning are as follows: Learning rates: {0.001, 0.0001}; Optimizers: {Adam, Stochastic Gradient Descent (SGD)}; and Regularization methods: {Batch Normalization, Dropout}.

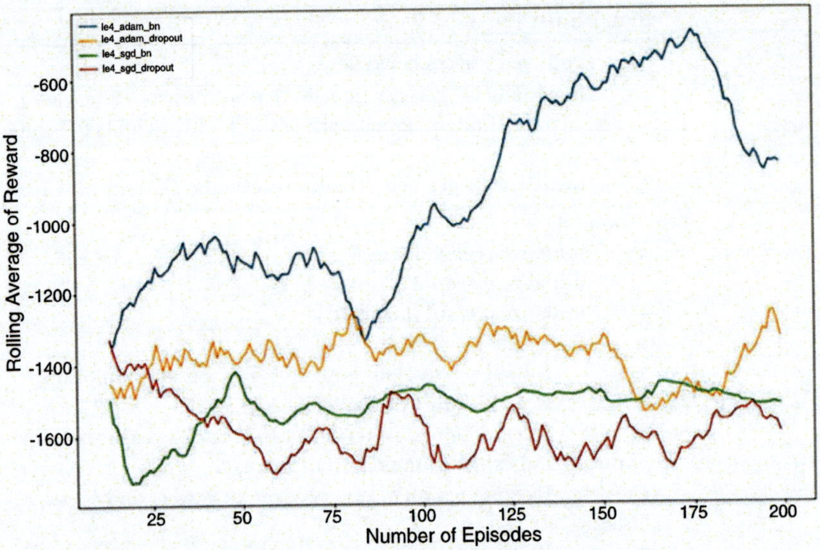


**Fig. 9.3** Rolling average of reward of all configurations trained normally

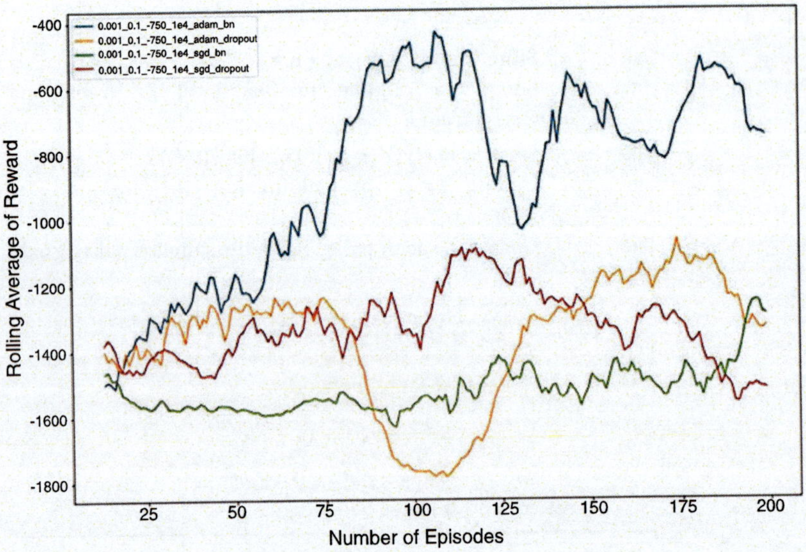


**Fig. 9.4** Rolling average of reward of all configurations trained using the proposed work

### 9.4.2 Evaluation Metric

We are deploying a basic metric for analyzing the improvement of our proposed work compared to other baseline models, and this measure that is used to evaluate each configuration is called the cumulative reward. It can be defined as the sum of the rewards the agent has earned throughout the episodes or training and testing.

### 9.4.3 Implementation Details

The relatively computationally less intensive environment "Pendulum-V1" from OpenAI's Gym library's 0.23.1 version was used to conduct our experiments. We used an NVIDIA RTX 3050 GPU equipped with the 11th generation Intel Core i7 processors on a system with 16 GB RAM to speed up the execution time. The implementation was done in Python using Keras and TensorFlow libraries using Visual Studio Code as IDE and Miniconda for Python (3.8.0) Environment. TensorFlow 2.8.0's Keras implementation was used to build and execute neural network operations, which used GPU power very efficiently.

### 9.4.4 Results

We ran the experiments with the configurations for a maximum of 200 episodes keeping all other settings as default. To clearly see the progression trends, we decided to smoothen the scores by applying a rolling average with a window of 14. Our results are presented in Figs. 9.3 and 9.4. Further, Fig. 9.3 is the performance on model trained normally, and Fig. 9.4 is trained with our proposed work. The best-performing configuration from the above experiments was trained using a better gamma value on a deeper self-attention-based network, and the observed cumulative rewards were plotted in the graphs presented in Figs. 9.5 and 9.6. Difference in performance can be clearly observed from the figures.

Figure 9.3 illustrates the change in rolling average of reward with increase in number of episodes for various configurations of DDPG model trained normally, whereas Fig. 9.4 represents change in rolling average of reward with number of episodes for the various configurations of DDPG model trained on the proposed algorithm. Figure 9.5 illustrates the change in rolling average of reward with increase in number of episodes for the best-performing configuration of DDPG model trained normally, whereas Fig. 9.6 represents change in rolling average of reward with number of episodes for the best-performing configuration of DDPG model trained on the proposed algorithm.

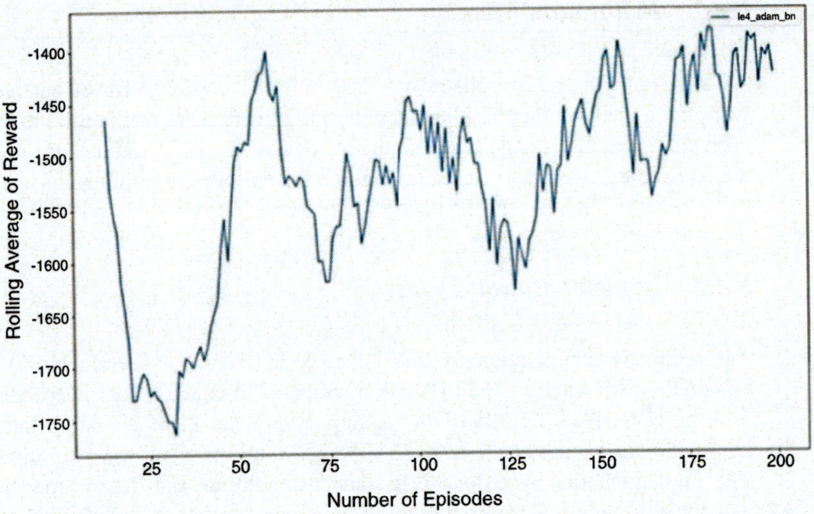


**Fig. 9.5** Cumulative reward of best performing configuration trained normally

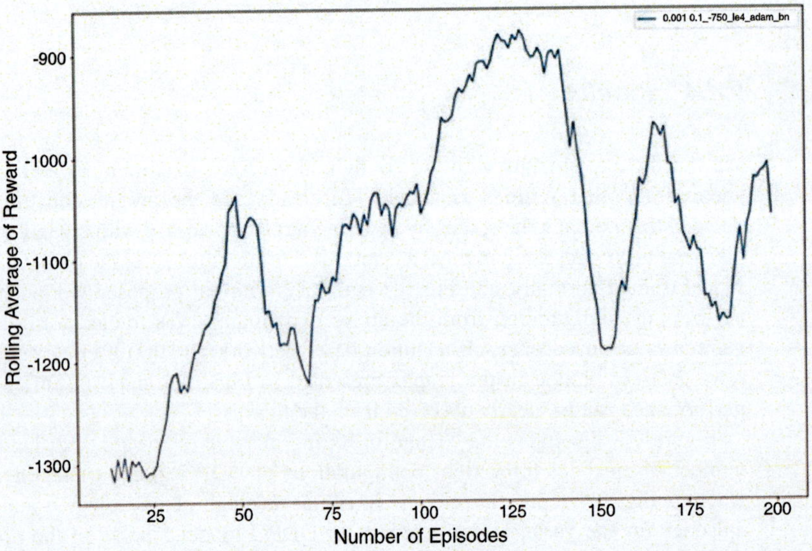


**Fig. 9.6** Rolling average of reward of best performing configuration trained using the proposed work

## 9.5 Conclusion and Future Scope

We tested the self-attention-based architecture for both the actor and the critic networks along with a self-attention architecture for the target actor and critic networks. It can be concluded that having self-attention in both pair of networks works better than the other network architectures. There is also considerable improvement observed due to the way the weights of the policy network are updated based on the updates of target network. The value of tau is changed based on reward earned, and this also suggests improvement in the policy learning. Another area we wish to explore in our future research projects would be applying the aforementioned approach to multiagent and noncontinuous environments.

## References

1. G. Dulac-Arnold, N. Levine, D.J. Mankowitz, J. Li, C. Paduraru, S. Gowal, T. Hester, Challenges of real-world reinforcement learning: Definitions, benchmarks and analysis. Mach. Learn. **110**(9), 2419–2468 (2021)
2. J. Wu, H. Li, Aggregated multi-deep deterministic policy gradient for self driving policy, in *International Conference on Internet of Vehicles*, (Springer, Cham, 2018), pp. 179–192
3. Y. Xiang, J. Wen, W. Luo, G. Xie, Research on collision-free control and simulation of single-agent based on an improved DDPG algorithm, in *2020 35th Youth Academic Annual Conference of Chinese Association of Automation (YAC)*, (IEEE, 2020), pp. 552–556
4. C. Kang, C. Rong, W. Ren, F. Huo, P. Liu, Deep deterministic policy gradient based on double network prioritized experience replay. IEEE Access **9**, 60296–60308 (2021)
5. J.C. Jesus, J.A. Bottega, M.A. Cuadros, D.F. Gamarra, Deep deterministic policy gradient for navigation of mobile robots in simulated environments, in *2019 19th International Conference on Advanced Robotics (ICAR)*, (IEEE, 2019), pp. 362–367
6. C. Liu, A.G. Lonsberry, M.J. Nandor, M.L. Audu, A.J. Lonsberry, R.D. Quinn, Implementation of deep deterministic policy gradients for controlling dynamic bipedal walking. Biomimetics **4**(1), 28 (2019)
7. C. Tao, J. Xue, Z. Zhang, Z. Gao, Parallel deep reinforcement learning method for gait control of biped robot. IEEE Trans. Circuits Syst. II Express Briefs **69**(6), 2802 (2022)
8. K. Hauser, *Hyperparameter Tuning for Reinforcement Learning with Bandits and Off-Policy Sampling*. Doctoral dissertation, Case Western Reserve University (2021)
9. N.M. Ashraf, R.R. Mostafa, R.H. Sakr, M.Z. Rashad, Optimizing hyperprameters of deep reinforcement learning for autonomous driving based on whale optimization algorithm. PLoS One **16**(6), e0252754 (2021)
10. M.P.K. Mishra, P. Prakash, Environment interaction of a bipedal robot using model-free control framework hybrid off-policy and on-policy reinforcement learning algorithm. Int. J. Innov. Sci. Res. Technol. **6**(11), 277–288 (2021)
11. A. Kumar, N. Paul, S.N. Omkar, Bipedal walking robot using deep deterministic policy gradient. arXiv preprint arXiv:1807.05924 (2018)
12. N. Ahmadi, D. Cabo, S. Jain, R. Kip, *Comparing Continuous Reinforcement Learning Algorithms Using OpenAI*. (GitHub, 2023), https://sudhakaranjain.github.io/pdf/reports/ML_report.pdf. Last visited 10 Jul 2023
13. D. Grbic, S. Risi, Towards continual reinforcement learning through evolutionary meta-learning, in *Proceedings of the Genetic and Evolutionary Computation Conference Companion*, (Association for Computing Machinery, New York, 2019), pp. 119–120

14. D. Ha, Reinforcement learning for improving agent design. Artif. Life **25**(4), 352–365 (2019)
15. J. Hou, H. Li, J. Hu, C. Zhao, Y. Guo, S. Li, Q. Pan, A review of the applications and hotspots of reinforcement learning, in *2017 IEEE International Conference on Unmanned Systems (ICUS)*, (IEEE, 2017), pp. 506–511
16. A. Kuznetsov, A. Grishin, A. Tsypin, A. Ashukha, D. Vetrov, Automating control of overestimation bias for continuous reinforcement learning. arXiv preprint arXiv:2110.13523 (2021)

# Chapter 10
# EBC: Encounter Buffer and Contact Duration-Based Routing Protocol in Opportunistic Networks

**Satya Jyoti Borah, Jagdeep Singh, and Tekcham Davidson Singh**

## 10.1 Introduction

Ad hoc network [1] is an example of peer-to-peer wireless network without the need for an access point. Communication can be established between two or more nodes using an internal communication medium, such as Bluetooth or Wi-Fi. Every node in a multihop communication system like MANET [2] is thought to be beneficial to one another. They accept responsibility for contributing to and directing all network traffic. They do, however, have significant restrictions that make it simple for attackers, such as nodes/path discovery and poor security against unwanted inbound connections. In this chain, Opportunistic network [3] is one of the recent developments of MANETs. OppNets and MANETs are very similar and very different from one another. OppNets adhere to the asynchronous mode of communication, while MANETs adhere to the synchronous mode. In MANETs, message communication occurs when a definite path between the sender and the receiver has been established. OppNets, on the other hand, does not start out by assuming that a direct link already connects the two nodes that are trying to interact. The pair of source and destination nodes may not be connected to the same network or in close proximity to one another. Here, communication of messages among nodes takes place using store-carry-forward technique. Whenever nodes come to their communication range, message exchange takes place, and

S. J. Borah · T. D. Singh
Department of Computer Science and Engineering, North Eastern Regional Institute of Science and Technology, NERIST, Nirjuli, Itanagar, India

J. Singh (✉)
Department of Computer Science and Engineering, Sant Longowal Institute of Engineering and Technology, Longowal, Punjab, India



I. Woungang, S. K. Dhurandher (eds.), *5th International Conference on Wireless, Intelligent and Distributed Environment for Communication*, Lecture Notes on Data Engineering and Communications Technologies 174,
https://doi.org/10.1007/978-3-031-33242-5_10

they store the message in their buffer until a suitable node is available to forward the message toward its destination. OppNet, on the other hand, offers a platform for devices to converse in the earlier stated class of difficult network conditions. In the network, packets are frequently delayed while they wait for a path to the destination to open up. As a result of the additional delay that occurs during message delivery, OppNets are also known as DTN (Delay Tolerant Networks) [4]. Nevertheless, some applications, like emailing, can handle longer delays. Once more, frequent connections and disconnections between nodes may cause some OppNets applications to function less well. It is an extension to current networks that requires less infrastructure, is more affordable, and can be set up more quickly. Forwarding of messages in OppNets is not an easy task. It is entirely distinct as compared to traditional methods of routing. The major goals in this situation are to anticipate a dependable packet delivery, although there is no fixed route between the source and the destination. In this type of network, the nodes are movable by nature, making it impossible to completely forecast the pathways for packet delivery. Therefore, traditional routing protocols are useless in these networks

This paper proposed a novel routing protocol in OppNets, called EBC, which is based on three parameters, namely, €, β, and λ. (€) The number of times a node encounters with other node is consider as encounter value, (β) buffer space of node, and (λ) contact duration of each node with other node. It is also used to identify a group of nodes from the source neighbor as the best forwarder for the message.

Section 10.2 briefly describes some of OppNet's standard routing protocols. Section 10.3 details the proposed EBC protocol. Section 10.4 shows the simulation environment and results. Section 10.5 summarizes this paper.

## 10.2 Related Work

In recent times, several routing methods were discussed for OppNets. Vahdat and Becker in [5], introduced the Translate routing procedure. They broadcast messages to nodes within the communication range of the source or intermediate node. Every node in the network repeats the same message patterns, records them in its cache, and keeps an overall summary vector table for all messages. Next, exchange of their summary vectors took place for those messages that are missing in their buffer. Similar to how all communications are sent around the network, they eventually arrive at their intended recipient as quickly as possible with the highest delivery probability. The fundamental disadvantage of the aforementioned protocol is that, in comparison to the other OppNet protocols, it consumes more bandwidth, uses larger buffers, and produces the most congestion. In "History Based Prediction for Routing" (HBPR) [6], Dhurandher et al. discussed how to exploit the node context information for message transmission. In this work, they use the node's present movement history to predict its next locations. The location history of a node is created based on the moving sequence of nodes for an interval of time. In the history table, they consider the period of time that is required by a node to come in contact with another node, and the movement and direction of the nodes are considered from the history table, which are the major node's context information that is taken for transmitting message. In their work, the whole network area is divided into a number of cells where nodes can move, and to identify a particular cell, a unique ID is used. Then, they mark the cell as a home location where a node visits most of the time. The entire HBPR protocol operates in three stages: (1) Identifying home location: This phase starts by flooding their home location information of all the node in the network to inform the other nodes, and all the nodes assume to follow the human mobility model. On their movement toward destination, nodes may visit some of the location many times and some of the location a few times or may not visit at all. Further, node's visiting information to a location is used to predict the nodes next location. (2) Message creation and home location update: This phase can be performed with the two steps. In the initial step, the destination ID is stored for the messages that are generated by some of the nodes. Next, the information of a node whose home location is changed during message communication is immediately flooded in the network to inform other nodes to update accordingly, which later on helps the nodes to maintain their behavior and relation; and finally. (3) Selecting Next hop: In the third stage of HBPR, it depends entirely on three parameters of the node: such as its stability during motion, what will be its direction in future motion, and the perpendicular distance and its line-of-sight between source and destination. By exploiting these parameters, a matrix called "utility matric" is determined for selecting the next best node that can forward the message. Finally, the message is transferred to the node that has a higher utility value compared to a prescribed threshold value. It has been observed that HBPR uses human mobility model for their efficient performance, but it failed to show its performance in other mobility models [7].

In [8], Lindgren et al. proposed the "Prophet routing protocol," which follows the idea of encounter frequency between two nodes in the past and the transitive property. The transitive property states that if a node visits a place or meets another node, a greater number of times in the past, there is a maximum chance that the node will encounter that particular node or that particular location during its future movement. On the basis of this, Prophet calculates a probability matrix called "delivery probability" for all the nodes before transmitting a message toward its destination. Then, the protocol forwards the message from a node to another node whose probability matric value is greater than maximum probability value to the destination. Prophet is a selective routing protocol, so it accumulates less overhead ratio and less bandwidth utilization in the network, but in terms of deliver probability and average latency its performance is poor.

In [11], Spyropoulos et al. discussed about the Spray-and-Wait routing protocol. This work constitutes two phases, that is, Spray phase and Wait phase. In the first phase of their scheme, the source node, instead of flooding all he messages into the network, selects only M number of neighboring nodes. Next, Wait phase is started, if M relay nodes never encounter the destination during Spray phase. In Wait phase, the relay nodes that carry the message copies transmit directly to the destination. As more message copies are sprayed in the Spray phase, at least a single copy of the message swiftly reaches the destination. This protocol is also one kind of dissemination type, but Spray phase controls the dissemination of the message within the network. Hence, the message delivery probability is high; however, network congestion is also high.

The Genetic algorithm-based scheme OppNets was put forth by Dhurandher et al. in [10]. In their method, communications are routed from source to destination using a Genetic Algorithm (GA). The details about the home of a node's groups and its neighbor's node groups are recorded on basis of the node's personal information table. Based on this, the source node uses GA to select the next step as follows. To find the instantaneous fit of a node for being a suitable forwarding node, a suitable function is developed, and a set of random chromosomes is initialized. By means of outcrossing and selection, a fresh set of chromosomes is produced. Until the best feasible optimization result is obtained, this process is repeated.

In [15], Dhurandher et al. suggested PRoWait routing for OppNets. The features of the routing protocols Spray and Wait [9, 10] and Prophet [8] were used to create it. When two nodes come together, a straightforward forwarding technique is typically used, and packets are exchanged from the node with the lowest probability of distribution to the node with the highest likelihood of dissemination of the point arrive. To choose a node as the next hop, the delivery predictability of a node is determined similarly to Prophet. On the other side, the Spray and Wait strategy is used to forward the messages to nearby nodes [12]. The packet is initially duplicated in n copies at the originating node. A node "x" (source/relay) that comes into contact with node "y" (having without copies) and has n > 1 copies of the packet transfers n/2 number of packets to "y" and retains n/2 packets for itself. In the event that node "y" comes across node "z" (with no copies), "y" gives "z" n/4 copies while keeping n/4 copies for him. This procedure keeps going till a node either ends up with a copy of the packet or reaches its destination. If the destination of the message is not known during this phase, each node holding a version of the message transmits directly to it. In this scheme, each nodes hold two properties: hop count and probability. When a node comes in to the contact with another node, a packet exchange takes place from a low probability to the one with a high probability; otherwise, no message transmission occurs. To ensure effective delivery to the intended recipient with the least amount of delay, PRoWait tries to keep several copies of the message. By minimizing the amount of flooding, it also seeks to conserve resources in comparison to the epidemic routing system.

In the literature as discussed above, many of the protocols designed for OppNets are based on dissemination-based routing protocols, which accumulate more network traffic due to maximum number of duplicate messages in the network as a result more network resources are required. On the other hand, context-based routing protocols fail to show their efficiency w.r.t. delay and "delivery probability" in data transmission. It is mainly the time required to compute the node's context information to select the next step for the message to the destination. In the above discussion, it has been observed that while developing a routing method for

OppNets, along with the probability of sending the message, the distance between the source node and the destination node must also be determined. Also, other parameter such as number of times a node encounters with target, node buffer space, duration of node exposure, and so forth cannot be used. Considering these limitations, an algorithm on the basis of encounter, buffer, and contact (EBC) for opportunistic networks is designed with an objective that messages can be delivered to their destination with minimum number of hop count, minimum delay, and maximum time the message remains in the nodes' cache.

## 10.3 System Model

The suggested technique makes use of *N* mobile nodes that are mutually beneficial in an opportunistic network setting; the nodes participate in this network are considered to have sufficient battery life, and no malicious activity was performed among them during message transmission. In this proposed scheme, the next best hop of a message to its destination is derived based on three parameters, namely, €, β, and λ. The encounter values of each node with other nodes in the network are originally generated dynamically by the EBC to obtain these parameters. The number of times a node encounter with other node is considered as encounter value, which is represented by €. Next the scheme calculates the buffer space of every node in the network and is denoted by β. Then EBC calculate the contact duration of each pair of nodes in the network. It is the number of times for which the nodes are interconnected with one another and denoted by λ which is given in Eq. 10.1 to Eq. 10.3.

Next, to obtain the next hop from the network by the source, the scheme calculates the average value of each parameters and store them in *getAvgEncounterFor, getAvgBufferFor*, and *getAvgContactDurationFor*. Then, the scheme compares each parameter, that is, €, β, and λ with this average calculated value; those nodes will select from the neighbor of the source whose corresponding values are greater than equal to the calculated average value of the three parameters and stores it in **hashmap1**, **hashmap2** and **hashmap3,** respectively.

Again, to find the best forwarding hop within the network of the message toward its destination, the scheme maximizes encounter value, buffer space, and contact duration and derives a new parameter called best forwarding parameter, which is denoted by τ as shown in Eq. 10.4. Likewise, EBC identifies the nodes from the networks that have Γ value, which is greater than or equal to a predefined threshold value and store all the nodes in hashmap-EBC. Finally, the proposed EBC identifies the nodes available in the hashmap-EBC as next best hop of the message whose Current-Energy value is greater than the Current-energy value of the source. The algorithm of the proposed EBC is presented in Algorithm 10.1 on page 139.

$$€ = Encounter = getEncounter(source, dest.) \tag{10.1}$$

$$\beta = Buffer = getFreeBufferSize(node)) \tag{10.2}$$

$$\lambda = \text{ContactDuration} = getConnectionTime\,(source, dest.) \tag{10.3}$$

$$\tau = \text{hashmap1} * \text{hashmap2} * \text{hashmap3} \tag{10.4}$$

## 10.4 Evaluation

In this section, the proposed model (EBC) is simulated using Opportunistic Network Environment tool [13] and compared the performance of EBC with two benchmark routing techniques HBPR and Epidemic.

- Movement of nodes can be analyzed using different mobility models [14].
- Data Packets can be routed among nodes with different oppNet routing protocols.
- With the help of GUI, mobility pattern and packet transmission can be visualized in real time.

In this simulation, a total of six groups of mobile nodes—three groups of pedestrians, three groups of trams, and three groups of buses—are taken into account to examine the performance of the suggested protocol. Walking pace for a group of 30 pedestrians ranges from 0.5 to 1.5 m per hour. Each of the two tram groups has two nodes and travels at a speed of 7–10 m/h. Each pedestrian node will receive 15 MB of buffer space, whereas each tram node will receive 50 MB. With a 20-m transmission range and a 250 Kbps transmission speed, Bluetooth serves as a communication medium. The high-speed interface has a 1500 m range and a 10 Mbps transmission speed. The network assigns 100 minutes of time-to-live (TTL) for each message generation. A fresh message with a size of 500 KB to 1 MB is created every 25–35 seconds. The simulation space is 4500 × 3400 square meters in size, and each simulation lasts 100,000 seconds. The "shortest path map-based movement model" tracks the movement of nodes [15]. We have altered the TTL from 100 to 300 minutes, the number of nodes from 66 to 186, and the buffer time average from 25 to 75 minutes, and studied the performance of EBC and compared it against that of Epidemic and the History-based Prediction for Routing (HBPR) routing protocols.

**Algorithm 10.1 Encounter buffer and contact duration-based routing protocol**

```
//getEncounter(source,dest.) returns the number of encounter
              value of source w.r.t. destination.
// getFreeBufferSize(nodes) returns the node's storage
              capacity.
// contactDuration(source,dest.) returns the source's contact
              duration w.r.t the destination.
// getCurrent-Energy(source) returns the energy of the
              source node.
// getCurrent-Energy (node in hashmap-EBC) returns the energy
              of nodes in hashmap-EBC.
Begin
     for each message do
           for each neighbour n of the source/interim.
              node do
                    € = getEncounter(source,dest.)
                      if € >=  getAvgEncounterFor(dest.)
                         Insert node in  hashmap1
                      end if
           β = getBuffer(n)
           if β >=  getAvgBufferFor(nodes)
               Insert node in  hashmap2
           end if
           λ = contactDuration(source,dest.)
           if λ >=  getAvgContactDurationFor(dest.)
               Insert node in  hashmap3
           end if
           τ= node from hashmap1* node from hashmap2
                      * node from hashmap3
           end for
        end for
        for each neighbour do
               if τ (of neighbour) ≥ T (threshold) then
                    Insert node in  hashmap-EBC
               end if
        end for
        for each node in  Energy-EBCdo
     if getCurrent-Energy (node in hashmap-EBC) >
          getCurrent-Energy(source) > then
        Insert node in  Energy-hashmap-EBC
        end if
        end for
        for each node in  Energy-hashmap-EBCdo
      Transfer message from source/intermediate
         node to nodes in  Energy-hashmap-EBC
               end for
End
```

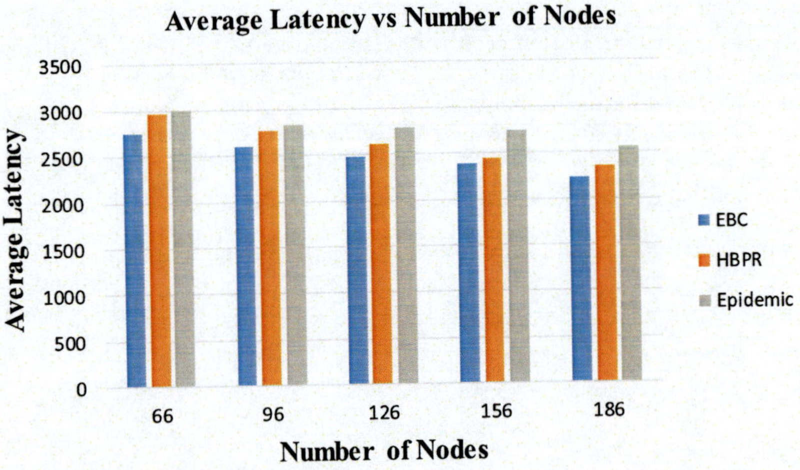


**Fig. 10.1** Average latency versus number of nodes

Figure 10.1 shows the relation between average latency and number of nodes. It reflects from Fig. 10.1 the varying number of nodes in the average latency for all the protocol decreases. This is because there are more nodes in the network, which means there are more carriers to carry the packets to its destination, which reduces the amount of time needed to send the message. While the number of nodes varies, the performance of the proposed EBC is 5.25% better than HBPR and roughly 10.77% better than Epidemic in terms of latency average.

Figure 10.2 is drawn between average latency and message interval; it has been found from the graph that when varying the message generation interval, the average latency is also decreasing for all the protocols. This is caused by an increase in the message creation interval, which leads to more messages being created in the network and, as described in Fig. 10.1, a decrease in average latency. After comparing the result obtained by EBC, w.r.t average latency is 8.55% superior as compared to HBPR and 8.44% superior than Epidemic routing protocol while varying message generation interval.

Figure 10.3 depicts the relation between average latency and TTL (time-to-live). It can be noticed from the graph that when varying the TTL, the average latency for all the protocols is increasing. The reason is that while increasing time-to-live of the message, most of the message reside in nodes buffer, and hence, more time is required to deliver the message to its destination and as a result average latency increase. From the result in Fig. 10.3, it has been found that the proposed protocol EBC outperforms in terms of average latency compared to HBPR and Epidemic. It is 6.93% better than HBPR and 11.88% better than Epidemic while varying the TTL.

Figure 10.4 shows the graph between hop count and number of nodes. From the graph it can be seen that the hop count for all the three protocols grows as the

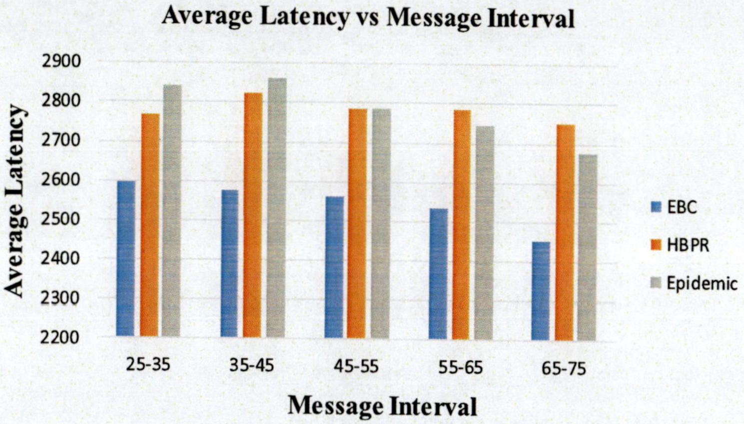


**Fig. 10.2** Average latency versus message generation interval

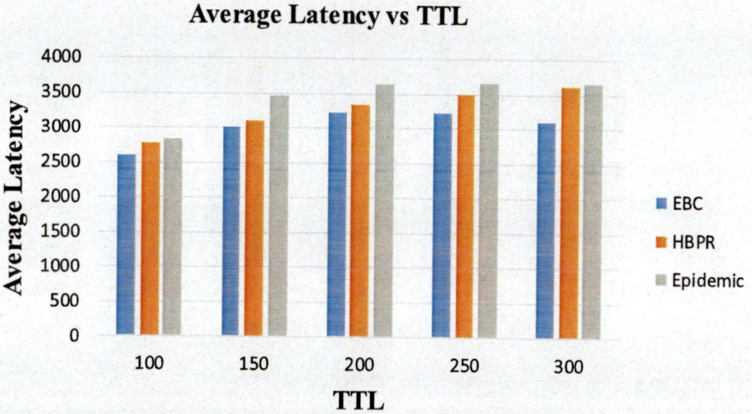


**Fig. 10.3** Average latency versus time-to-live (TTL)

node changes from smaller numbers to higher numbers. This is because there are more intermediate nodes involved in sending the message to its destination as there are more nodes in the network, which increases the number of hops. The proposed EBC outperforms in terms of hop count as compare to HBPR and Epidemic routing protocol. The performance of EBC protocol is 6.76% better than HBPR and 12.67% better than Epidemic protocol when number of nodes is varying.

Figure 10.5 is drawn between hop count and message creation interval. It can be seen from the graph that when the message generation interval varies, the hop count also increases for all the protocols. This is due to an increase in message creation intervals, more messages being formed in the network, and more nodes being required to send each message to its destination, leading to an increase in the

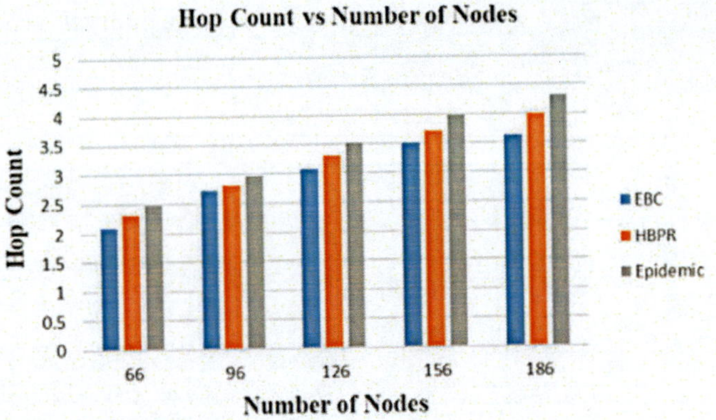


**Fig. 10.4** Hop count versus number of nodes

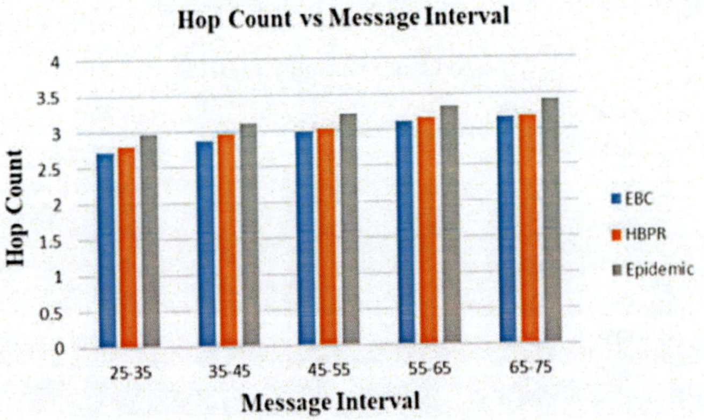


**Fig. 10.5** Hop count versus message generation interval

number of hops. After comparing, the result obtained by EBC is 2.13% superior than HBPR, whereas result is around 7.45% superior than the Epidemic protocol in terms of hop count while varying message creation interval.

Figure 10.6 depicts the relation between hop count and TTL (time-to-live). It can be found from the Fig. that when varying the TTL, the hop count for all the protocols is increasing. The reason is that while increasing TTL of the message, most of the message reside in nodes buffer due to increases the live of the messages and hence more number of nodes involve in message transmission to its destination. Hence, the hop count increases. From the result in Fig. 10.6, it has been seen that the performance of EBC in terms of hop count is around 4.21% better than HBPR and around 12.73% better than the Epidemic protocol while varying the TTL.

Figure 10.7 plots the graph between buffer time average and number of nodes. From Fig. 10.7, it is observed that, by varying number of nodes the buffer time

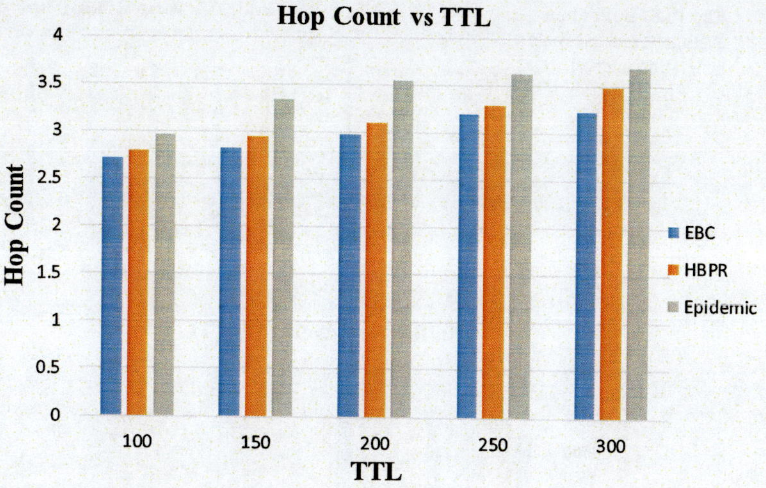


**Fig. 10.6** Hop count versus time-to-live (TTL)

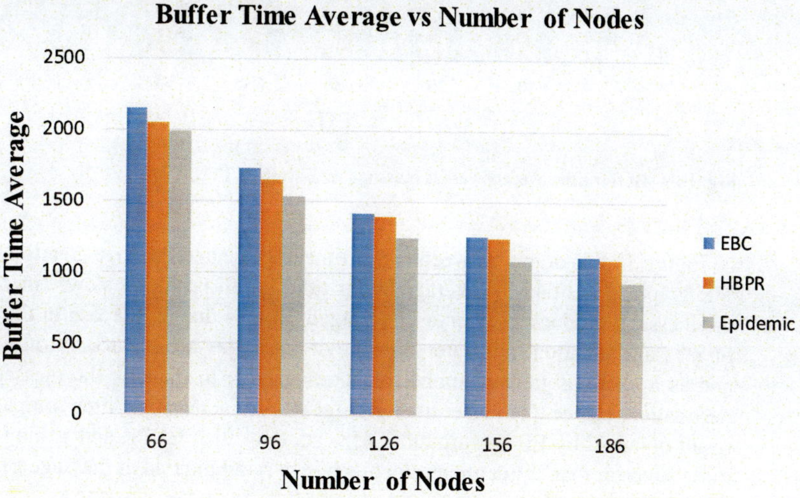


**Fig. 10.7** Buffer time average versus number of nodes

average decreases for all protocols. This is because, as the number of nodes rises, more buffer space for messages in the network becomes available. As a result, fewer messages are stored in nodes' buffers, which lowers the average buffer duration. The performance of proposed EBC is 2.88% better than HBPR and 13.75% better than Epidemic protocol in terms of buffer time average when the number of nodes varies.

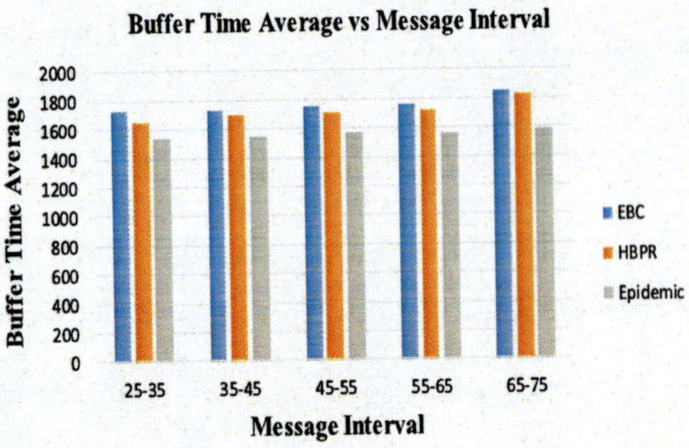


**Fig. 10.8** Buffer time average versus message generation interval

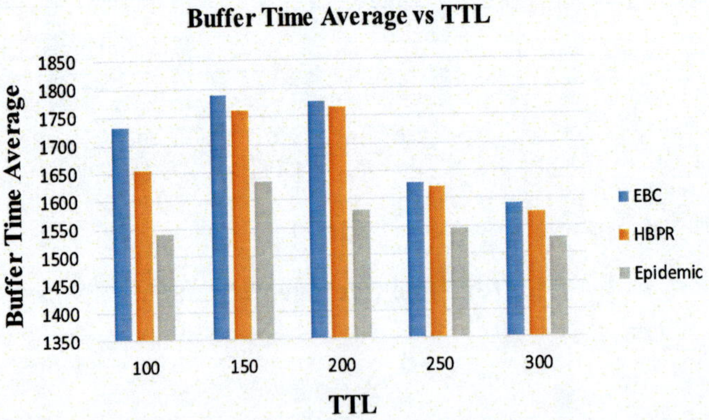


**Fig. 10.9** Buffer time average versus message time-to-live (TTL)

Figure 10.5 is drawn between buffer time average and message creation interval. The graph shows that the average buffer time for all protocols slowly increases as the message production interval is changed. This is due to increase in the interval of message creation; a greater number of messages are created in the network, and as a result, a greater number of nodes involve to transmit the message to its destination. Hence, the buffer time average is also increasing. After comparing, the result obtained by EBC is around 2.45% superior than HBPR and around 13.47% more superior than Epidemic protocol while varying interval of message generation (Fig. 10.8).

Figure 10.9 depicts the relation between buffer time average and TTL (time-to-live). It can be found from the graph that when varying the TTL, the buffer time average for all the protocols deceases. After calculating the result of EBC and comparing with HBPR and Epidemic protocol, the proposed EBC outperforms in terms of buffer time average. The performance of EBC in terms of buffer time average is around 1.61% better than HBPR and around 8.74% better than Epidemic protocol while varying the TTL.

## 10.5 Conclusion

The EBC protocol, developed on the basis of the surrounding node's encounter value in addition to its buffer capacity and contact time, is proposed. The work focuses in the direction of enhancing the performance of EBC by applying three parameters, namely, €, β, and λ, to identify the nodes among the neighboring nodes that are selected as good forwarders from the EBC protocol. According to simulation results, the suggested EBC performs better than the established protocols HBPR and Epidemic over the measuring parameters average latency, hop count, and buffer time average. The said scheme is unable to dominate the measuring parameters like message delivery probability, message dropped, and overhead ratio. The energy and security problems with the proposed method can be resolved in the future. Additionally, the proposed approach can be tested with other mobility models and circumstances.

## References


1. E.M. Royer, C.K. Toh, A review of current routing protocols for ad hoc mobile wireless networks. IEEE Pers. Commun. **6**(2), 46–55 (1999)
2. C.-K. Toh, *Ad Hoc Mobile Wireless Networks: Protocols and Systems* (Prentice Hall PTR, Englewood Cliffs, 2002)
3. L. Pelusi, A. Passarella, M. Conti, Opportunistic networking: Data forwarding in disconnected mobile ad hoc networks. IEEE Commun. Mag. **44**(11), 134–141 (2006)
4. K. Fall, A delay-tolerant network architecture for challenged internets, in *Proceedings of the 2003 Conference on Applications, Technologies, Architectures, and Protocols for Computer Communications*, (ACM Digital Library, New York, 2003), pp. 27–34
5. A. Vahdat, D. Becker, *Epidemic Routing for Partially-Connected Ad Hoc Networks, Technical Report CS-2000-06* (Department of Computer Science, Duke University, Durham, 2000)
6. S.K. Dhurandher, D.K. Sharma, I. Woungang, S. Bhati, HBPR: History based prediction for routing in infrastructure-less opportunistic networks, in *Proc. of 27th IEEE AINA 2013*, Barcelona, Spain, pp. 931–936
7. S.K. Dhurandher, D.K. Sharma, I. Woungang, Mobility model-based performance evaluation of the history based prediction for routing protocol for infrastructure-less opportunistic networks, in *Proc. of 10th IEEE,MOBIQUITOUS 2013*, Tokyo, Japan, 2–4 Dec 2013, pp. 757–767
8. A. Lindgren, A. Doria, O. Schelen, Probabilistic routing in intermittently connected networks. ACM SIGMOBILE, Mob. Comput. Commun. Rev. **7**(3), 19–20 (2003)
9. S.K. Dhurandher, J. Singh, P. Nicopolitidis, R. Kumar, G. Gupta, A blockchain-based secure routing protocol for opportunistic networks. J. Ambient Intell. Humaniz. Comput. **13**(4), 2191–2203 (2022)
10. N. Gupta, J. Singh, S.K. Dhurandher, Z. Han, Contract theory based incentive design mechanism for opportunistic IoT networks. IEEE Internet Things, 1–11 (2021). https://doi.org/10.1109/JIOT.2021.3109162
11. T. Spyropoulos, K. Psounis, C.S. Raghavendra, Spray and wait: An efficient routing scheme for intermittently connected mobile networks, in *Proc. of SIGCOMM Workshop on Delay-Tolerant Networking*, PA, USA, 22–26 Aug 2005, pp. 252–259
12. S.K. Dhurandher, D.K. Sharma, I. Woungang, R. Gupta, S. Gupta, GAER: Genetic algorithm based energy-efficient routing protocol for infrastructure-less opportunistic networks. J. Supercomput., Springer **69**(3), 1183–1214 (2014)

13. A. Keranen, J. Andott, *Opportunistic Increasing Reality for DTN Protocol Simulations, Special Technical Report* (Helsinki University of Technology, Networking Laboratory, 2007)
14. A. Keranen, J. Andott, *Opportunistic Increasing Reality for DTN Protocol Simulations*. Special Technical Report (2007)
15. S.K. Dhurandher, S.J. Borah, M.S. Obaidat, D.K. Sharma, S. Gupta, B. Baruah, Probability-based controlled flooding in opportunistic networks, in *2015 12th International Joint Conference on e-Business and Telecommunications (ICETE)*, vol. 6, (IEEE, 2015), pp. 3–8

# Chapter 11
# Socioeconomic Inequality Exacerbated by COVID-19: A Multiple Regression Analysis with Income, Healthcare Insurance, and Mask Use

**Wei Lu and Vanessa Berrill**

## 11.1 Introduction

COVID-19 is the largest pandemic in a century and has been widely recognized as the largest crisis in public health and economics since the Great Depression. In particular, The United States has been significantly hit compared to other countries, accounting for more than 15% of COVID-19 cases and deaths worldwide as illustrated in Fig. 11.1 [1].

Recent studies by Dr. Joseph S. Vavra, Professor of Economics at Chicago Booth, stated that "the initial economic hardship the virus has created has fallen primarily on the low-skilled and low-income part of the workforce, thus having a significant potential to exacerbate the economic inequality" [2]. Figure 11.2 also illustrates that the total number of COVID-19 cases and deaths is geographically disproportionate. This is mainly because disadvantaged low-income groups are more likely to need to stay in the workforce to support their living, and the work they do are usually public-facing jobs such as elder care and cleaning services, therefore increasing the risk of their exposure to COVID-19.

To address this issue, in this paper, we investigate two socioeconomic questions related to COVID-19 using data visualization, namely, (1) is there any relationship between median household income in a region and the number of COVID-19 cases and deaths in that region, and (2) is there any relationship between healthcare insurance coverage rate in a region and the number of COVID-19 cases and deaths in that region. The visualization results using the data collected from US Census

W. Lu (✉) · V. Berrill
Department of Computer Science, Keene State College, The University System of New Hampshire, Keene, NH, USA
e-mail: wlu@usnh.edu



I. Woungang, S. K. Dhurandher (eds.), *5th International Conference on Wireless, Intelligent and Distributed Environment for Communication*, Lecture Notes on Data Engineering and Communications Technologies 174,
https://doi.org/10.1007/978-3-031-33242-5_11

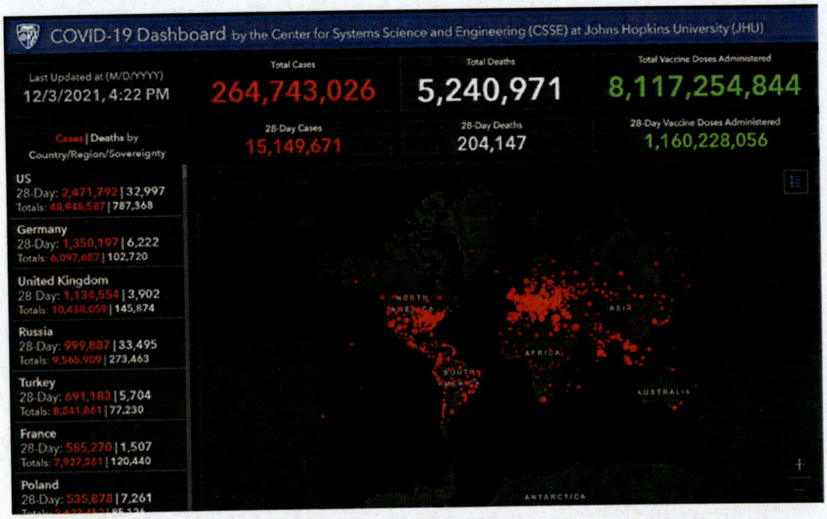


**Fig. 11.1** The no. 1 ranked total number of COVID-19 cases and deaths as of 12/3/2021 in the United States

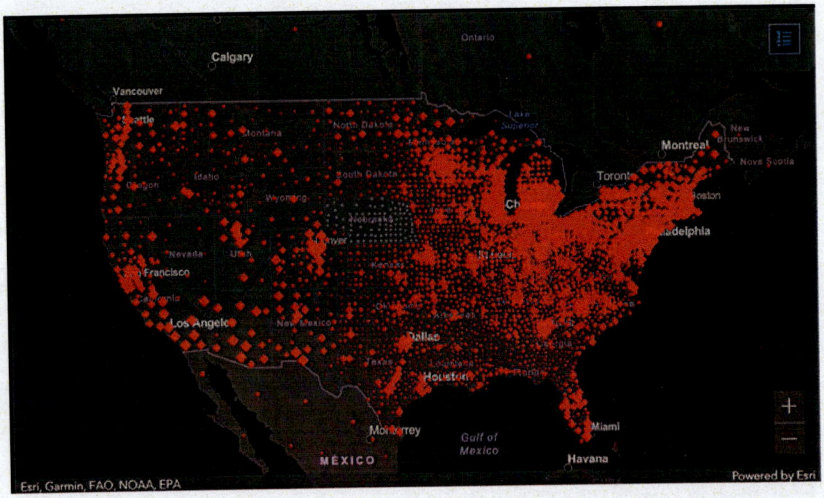


**Fig. 11.2** The total number of COVID-19 cases and deaths in the United States as of 12/3/2021

and New York Times illustrate that the poorer and more vulnerable social groups in terms of income and healthcare coverage have a disproportionately higher number of cases and deaths caused by COVID-19, validating that an increase in income and health insurance coverage inequality contributes to an increase of COVID-19 incidences. To help mitigate the impact of this socioeconomic equality, we also investigate the third question in this study, that is, does the use of masks help to reduce the number of COVID-19 cases and deaths? The visualization results using the dataset collected from the New York Times illustrate the effectiveness of using masks to help mitigate the spread of COVID-19 diseases. This is also reflected in the multiple linear regression model including the number of COVID-19 cases, the household median income, mask use, and health insurance by county. There are some outliers, though, where both the percentage of people always wearing masks and COVID-19 cases are very high; this is most likely because (1) a high case rate pushes more people to wear masks, and (2) those outliers simply represent network anomalies generated during data collection [3–6]. As a result, we conclude that wearing the mask is not only extremely important for high-risk groups for COVID but could also be one of the most effective solutions to overcome immediately the socioeconomic inequality before decision-makers propose more strengthening strategies to help relieve the pain of disadvantaged and minority groups impacted severely by COVID-19.

## 11.2 Related Works

Studies on the socioeconomic aspects and their influence on the COVID-19 infection rate during the pandemic are new. In [7], Hawkins studied the difference of rates in Massachusetts by the social determinates of health using three models to analyze the factors of the socioeconomic aspects, including median income; the percentage of uninsured, below the poverty line, and unemployed residents; and percentage of workers employed in the transportation and healthcare and social assistance industries. This study concluded that populations of people living in poverty; workers in healthcare and social assistance, transportation industries, service, and healthcare support positions; and renters, uninsured, and unemployed are more likely to have higher positive cases. Similar works are also done in [8, 9].

In [8], Krieger et al. analyzed the Commonwealth of Massachusetts COVID-19 death rates by zip code and social metrics, including poverty line, household crowding, percentage of people of color, and the Index of Concentration at the Extremes for racialized economic segregation. The analysis reports showed that there were 7016 deaths, more from January 1, 2020, through May 19, 2020, than the average should have been during this time. The results were gone through all the zip codes, but the zip codes in the top two poverty and household crowding categories were the most negatively impacted.

In [9], Liao and De Maio introduced how the COVID-19 pandemic has disproportionately impacted minoritized populations. Age-adjusted death rates in the African-American population are at 3.6 higher, and the Hispanic population is at 3.2 higher compared to the White population. These statistics represent about 30,000 more deaths. A variable that can explain the disproportionate impacts would be income inequality as well as politics and the relationship to the population's health. The data in this paper is collected from 3142 counties in the 50 states and includes Washington, DC. The main sources for this data are the Centers for Disease Control and Prevention, USAFacts.org, the US Census Bureau, the American Community Survey, GitHub, the Kaiser Family Foundation, the Council of State Governments, and the National Governors Association. There were political variables like if the governor was facing a term limit, was Republican or was male, and whether or not a Medicaid expansion was acted upon through the Affordable Care Act (ACA). There were also structural variables like using percentages of the African-American and Hispanic populations, using the Gini index, and political affiliations to associate with health. The results that were reported were that both outcomes were skewed to the right and were a negative binomial distribution. The paper concluded that the findings of this study show that it is important for government and health officials to understand these inequalities to be able to do their best to close that gap.

Several other works related are mainly focused on inequalities within different races concerning COVID-19. In [10], Bello-Chavolla et al. investigated the population of Mexico and evaluated why the older population has disproportionate COVID-19 cases compared to the younger population. Instead of accepting the idea of old age being the only cause of higher COVID-19 cases, Bello-Chavolla et al. in [10] looked at comorbidity and pointed out inequalities of the population of Mexico to develop comorbidities at a higher rate. The data sources for this analysis are from the General Directorate of Epidemiology of the Mexican Ministry of Health. Mixed-effects models were used because Mexico has very large inequalities when it comes to healthcare. Cox proportional hazard regression models were also used, as well as the Newton–Raphson algorithm to show the mortality rates. Using this regression model, pneumonia was looked at to see if the rise in pneumonia symptoms could identify a COVID-19 case, as well as looking at age, structural factors, or the number of comorbidities. Experimental results showed that adults with comorbidity had an increased risk of being put on ventilation and being in the ICU with COVID-19 and had a higher prediction of mortality. Another finding was that there was a higher risk for COVID-19 in metropolitan areas. Overall, it was found that older adults without comorbidity did have lower mortality rates compared to older adults with comorbidity. The proposed solution would be to evaluate the older population not just by age but also by comorbidities, functional status, and other identifiers of biological aging and work toward eliminating socioeconomic inequalities.

In [11], Kim and Bostwick analyzed the inequalities within different races concerning the COVID-19 pandemic by looking at particularly the population of African-Americans in Chicago. A social vulnerability index (SVI) was created using the sociodemographic characteristics of the 77 Chicago Community Areas (CCA) to determine social vulnerability. Data collected from the US Census American Community Survey included percentages of poverty, percentages of less than high school education, percentages of female-headed households with children, median household income, and employment ratio. Health risks were scored, including rates of heart-related death, stroke deaths, asthma, hypertension, diabetes, obesity, and smoking. Overall, the communities in Chicago that have a majority African-American population have a higher risk for COVID-19 cases and deaths. There are factors, such as poverty, segregation, and discrimination, which impact the African-American community's risk of COVID-19 and the ability to survive throughout the COVID-19 pandemic. Similarly, in [12], Millett et al. assessed the disproportionate impact of COVID-19 on African-American communities. Counties that had more than 5% of African-American residents were found to have higher rates of COVID-19 diagnoses. Tai et al. in [13] studied the disproportional impacts of medical, social, economic, environmental, and political contexts that have been a disproportional issue before the pandemic started but have worsened over time during COVID-19. One of the important statistics highlighted in [13] is that the mortality rate from COVID-19 in African-Americans is two times higher than the white population, stemming from chronic preexisting medical conditions and thus making the COVID-19 pandemic harder on the minorities. The paper also shows disparities in working from home, which helps avoid contact with COVID-19, where only 20% of African-American workers were able to work from home, and in New York City, 75% of the frontline workers are people of color. When looking at disparities in living conditions, there is a higher risk for COVID-19 in communities of people of color due to the density of people in the community.

In [14], Oronce et al. collected data reported from New York and Chicago that African-Americans and Hispanic people have been disproportionately impacted by COVID-19 cases and deaths because of discrepancies in economic segregation, decreased social mobility, and inaccessibility to healthcare. Also, income inequality is a large discrepancy where lower-income people are more likely to be essential workers who will get exposed at a larger rate to COVID-19. The paper looks at the relationship between income inequality and the number of COVID-19 cases and deaths using the Gini index and data from the 2018 American Community Survey. For the COVID-19 data, a dataset from the Center for Systems Science and Engineering from Johns Hopkins University was used. The data comes from the 50 states and is in the time frame of January 22, 2020 until April 13, 2020. The Spearman rank-order correlation test was used to get a correlation analysis. Multivariable regression was used as well. Variables in this analysis included the proportion of the population that was 65 years or older, female, African-American, Hispanic, below poverty, the median household income, tests performed per capita, doctors per capita, beds per capita, and stay-at-home or shelter-in-place orders. The paper reports a positive correlation between the Gini index and the rate of COVID-19 cases and deaths. This means states that have a higher Gini index will experience more deaths from COVID-19.

## 11.3 Datasets on COVID-19, Mask Use, Income, and Insurance

In this paper, we address COVID-19 cases and deaths and their associations with mask use, health insurance coverage, and median household incomes. In particular, the dataset includes four CSV files: (1) *us-counties.csv* is the county-level data for COVID-19 cases and deaths that includes seven-day averages and per 100,000 counts; (2) *mask-use-by-county.csv* is an estimated data regarding mask use by county in the United States; (3) *median-household-income.csv* is the median income data collected in the past 12 months of 2019 based on the inflation-adjusted dollars; and (4) *health-insurance-coverage.csv* is the data regarding the health insurance without coverage status by county in the United States collected in the past 12 months of 2019.

The first two data files (i.e., *us-counties.csv* and *mask-use-by-county.csv*) are downloaded from https://github.com/nytimes/covid-19-data, and the last two data files (*median-household-income.csv*, *health-insurance-coverage.csv*) are downloaded from https://data.census.gov/

New York Times (NYT) collects *us-counties.csv*, *mask-use-by-county.csv*, and The US Census Bureau collects *median-household-income.csv* and *health-insurance-coverage.csv*. In particular, the *us-counties.csv* was collected starting from the first reported coronavirus case in Washington State on January 21, 2020 and then tracked cases of coronavirus in real-time as they were identified after testing. Data in the *mask-use-by-county.csv* was collected based on about 250,000 survey responses between July 2 and July 14, 2020. The 2019 data on household median income and health insurance coverage status included in the two files *median-household-income.csv* and *health-insurance-coverage.csv* was first released on September 17 2020 by US Census.

The data in *us-counties.csv* was collected actively by the NYT team from databases operated by state and local governments and health departments. The data in *mask-use-by-county.csv* was collected "from a large number of interviews conducted online by the global data and survey firm Dynata at the request of The New York Times." "The firm asked a question about mask use to obtain 250,000 survey responses between July 2 and July 14, enough data to provide estimates more detailed than the state level" [15]. The data in *median-household-income.csv* and *health-insurance-coverage.*csv was collected through American Community Survey (ACS), which is mandatory and governed by Federal laws.

In *us-counties.csv*, there are six fields, that is, date, county, state, fips (i.e., county code), cases, and deaths; for example, given a record in us-counties.csv, that is, 2020-01-21, Snohomish, Washington, 53061, 1, 0, it means that in the county Snohomish of Washington State, there was one COVID-19 case and zero COVID-19 death on January 21, 2020.

In *mask-use-by-county.csv*, there are six fields, that is, countyfp (i.e., county code), never, rarely, sometimes, frequently, always, for the example given a record in *mask-use-by-county.csv*, that is, 01001, 0.053, 0.074, 0.134, 0.295, 0.444, it means that in the county identified by code 01001 (i.e., county Autauga of Alabama), it is estimated that 5.3% of people will never wear the mask, 7.4% rarely use the mask, 13.4 sometimes wear a mask, 29.5% frequently wear the mask, and 44.4% always wear the mask.

There are 242 fields in *median-household-income.csv*, and the fields we will be using are GEO_ID, Name, and Median Household Income (dollars). GEOIDs are "numeric codes that uniquely identify all administrative/legal and statistical geographic areas for which the Census Bureau tabulates data," Name is the string of county name and state name, for example, Baldwin County, Alabama, and Median Household Income indicates the Median household income.

There are 116 fields in *health-insurance-coverage.csv*, and the fields we will be using are GEO_ID, Name, and No insurance rate. GEOIDs are "numeric codes that uniquely identify all administrative/legal and statistical geographic areas for which the Census Bureau tabulates data"; Name is the string of county name and state name, for example, Baldwin County, Alabama; and No insurance rate indicates the percentage of people without health insurance coverage in a particular county.

## 11.4 Data Visualization

We use Tableau Prep Builder to connect to our four datasets. For each dataset file, there are no potential problems, that is, there are not too many null values and distributions of data look good. During the cleaning work phase, we modify GEO_ID into FIPS code with five- or six-digit code, so we can join the tables together. After dataset wrangling, we have a combined dataset as illustrated in the following Fig. 11.3. The full dataset can be found in [16].

Figure 11.4 visualizes the data on median household income in a region and the number of COVID-19 cases and deaths in that region, which shows a trend that the higher the median household income, the lower the number of COVID-19 deaths per one thousand people in that region. As illustrated in Fig. 11.5, Loudoun County, VA, has the highest income of $151,800; its COVID-19 death is 0.463 per one thousand. Hartford County, CT, has the second-highest income of $138, 500; its COVID-19 death is 0.65 per one thousand. York County, ME, has the third-highest COVID-19 death rate of 5.529 per one thousand, and the county's income is $30,380, one-fifth of the highest household income; the same applied to Blount County, AL, which has the second-highest death rate of 6 per one thousand, and its income is $37,153, close to one-fifth of the highest household income at Loudoun County, VA.

Figure 11.5 visualizes the data on the healthcare insurance coverage rate in a region and the number of COVID-19 cases and deaths in that region, which shows a trend that the higher the percentage of people without healthcare insurance coverage, the higher the number of COVID-19 deaths per one thousand people in that region.

| | A | B | C | D | E | F | G | H | I | J |
|---|---|---|---|---|---|---|---|---|---|---|
| 1 | County Name | Countyfp | State Name | Cases | Deaths | Population 2020 | CasesRate | DeathRate | CaseRatePerThousand | DeathRatePerThousand |
| 2 | Baldwin County | 1003 | Alabama | 19392 | 262 | 231767 | 0.083670238 | 0.001130446 | 83.67023778 | 1.130445663 |
| 3 | Calhoun County | 1015 | Alabama | 12915 | 278 | 116441 | 0.11091454 | 0.002387475 | 110.9145404 | 2.387475202 |
| 4 | Cullman County | 1043 | Alabama | 8822 | 175 | 87866 | 0.100402886 | 0.001991669 | 100.4028862 | 1.991669133 |
| 5 | DeKalb County | 1049 | Alabama | 8399 | 173 | 71608 | 0.117291364 | 0.002415931 | 117.2913641 | 2.415931181 |
| 6 | Elmore County | 1051 | Alabama | 9228 | 178 | 87977 | 0.104891051 | 0.002023256 | 104.8910511 | 2.023256078 |
| 7 | Etowah County | 1055 | Alabama | 13042 | 307 | 103436 | 0.126087629 | 0.002968019 | 126.0876291 | 2.968018872 |
| 8 | Houston County | 1069 | Alabama | 9991 | 255 | 107202 | 0.093197888 | 0.002378687 | 93.1978881 | 2.378686965 |

**Fig. 11.3** An example of data entries included in the joined dataset after wrangling

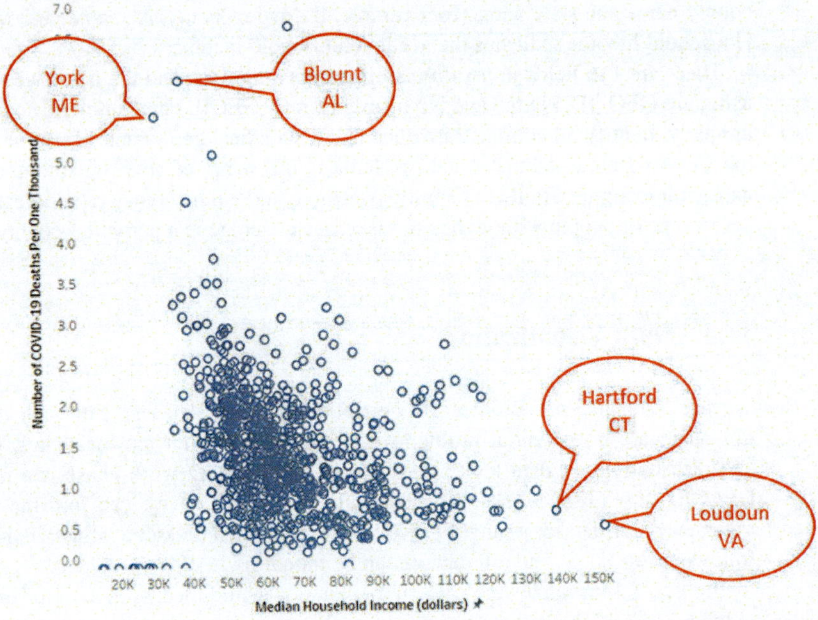


**Fig. 11.4** Median household income versus number of COVID-19 deaths per one thousand people

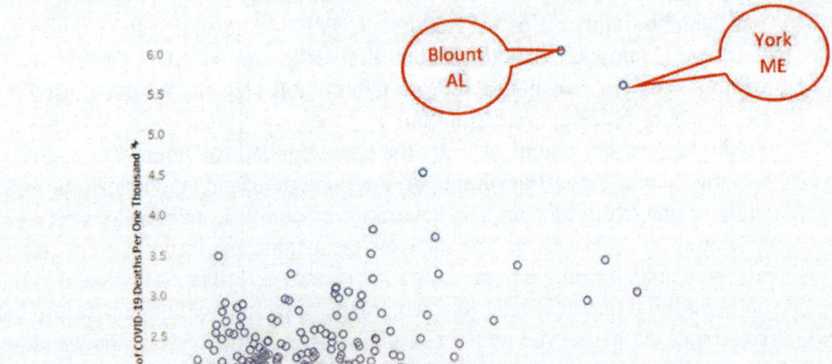


**Fig. 11.5** Healthcare insurance coverage rate versus number of COVID-19 deaths per one thousand people

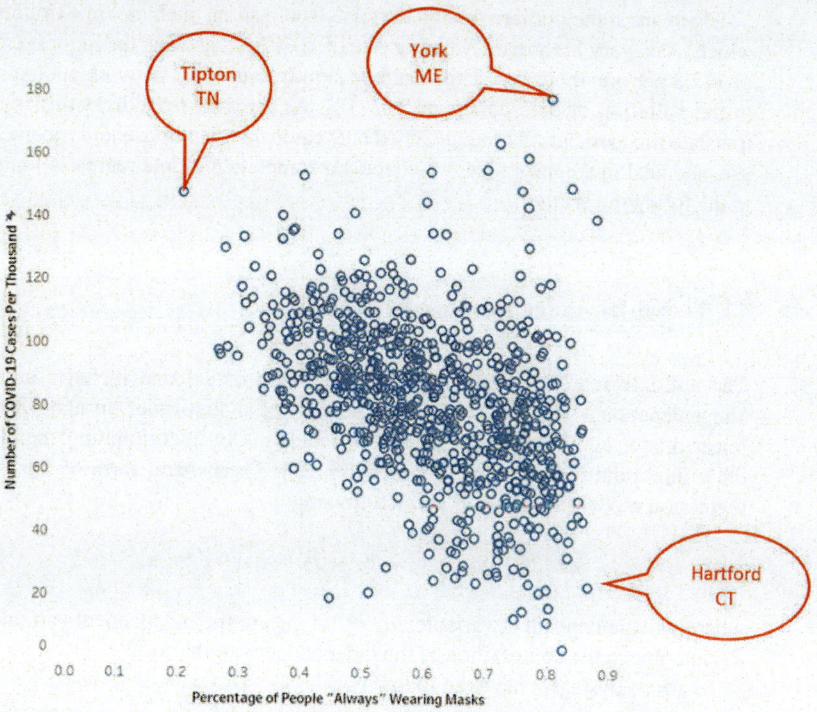


**Fig. 11.6** People who "always" wear masks versus number of COVID-19 cases per one thousand people

As illustrated in Fig. 11.5, Blount County, AL, has the highest death rate of 6 per one thousand, and it has about 27.1% of people who have no healthcare insurance coverage in this county; York County, ME, has the second-highest COVID-19 death rate of 5.529 per one thousand, and in this county, about 31.44% of people have no healthcare insurance coverage. On the other hand, Stokes County, NC, has a low uninsured rate of 3.06%, and its COVID-19 death is 0.374 per one thousand; the same applies to Clinton County, NY which has an uninsured rate of 3.28% with a 0.313 death per thousand.

Figure 11.6 visualizes the data on the percentage of people who always wear masks in a region and the number of COVID-19 cases and deaths in that region, which shows a trend that the higher the percentage of people who always wear masks, the lower the number of COVID-19 cases per one thousand people in that region. As illustrated in Fig. 11.6, Hartford County, CT, has the second-highest percentage of people (86.5%) who always wear masks in this county, and it has a very low COVID-19 case rate of 22.5 per one thousand; on the other hand, Tipton County, TN, with a lowest 20.7% rate of people using masks has a very high number of cases 147.7 per one thousand.

There are some outliers though, in this visualization, such as York County, ME, which has a very high mask-wearing rate of 80.4%; it also has the highest case rate of 177.8 per one thousand. This could be simply caused by network anomalies due to the pollution of data collection [17, 18]. As a result, to address this issue and translate the association among COVID-19 cases, median household income, mask use, and healthcare insurance, we conduct a complete multiple regression modeling in the following section.

## 11.5 Multiple Regression Modeling

The multiple regression model is a probabilistic model that includes more than one independent variable. It is an extension of a first-order straight-line-based linear model aiming to make accurate predictions by incorporating more potentially important independent variables [19, 20]. The general form of the multiple regression model is illustrated in the following:

$$y = \beta_0 + \beta_1 \times x_1 + \beta_2 \times x_2 + \cdots + \beta_k \times x_k + \varepsilon$$

where $y$ is the dependent variable; $x_1$, $x_2$, $\ldots x_k$ are the independent variables; and $\beta_k$ determines the contribution of the independent variable $x_k$.

In our analysis, we have the following:

$x_1$ = the percetage of people who always wear masks in a county
$x_2$ = the percetage of number of people who has no healthcare insurance
$x_3$ = the median household income in a county
$y$ = the number of COVID caese per one thousand in a county

The visualization results show that the number of COVID-19 cases and deaths depends on the people's behavior on mask use, their economic level (e.g., income), and their healthcare insurance coverage. Therefore, we can hypothesize the regression model in the following:

$$y = \beta_0 + \beta_1 \times x_1 + \beta_2 \times x_2 + \beta_3 \times x_3 + \varepsilon$$

where $y, x_1, x_2, x_3$ are illustrated above, and $\beta_1$, $\beta_2$, $\beta_3$ determines the contribution of the independent variable $x_1, x_2, x_3$, respectively.

$$E(y) = \beta_0 + \beta_1 \times x_1 + \beta_2 \times x_2 + \beta_3 \times x_3$$

and it is the deterministic portion of the regression model, while $\varepsilon$ is a random error component. The modeling with least-squares gives the parameter estimates as illustrated in Table 11.1 where can find that:

**Table 11.1** Parameter estimates

| Term | Estimate | Std. error | *t* Ratio | Prob > \|*t*\| | Lower 95% | Upper 95% |
|---|---|---|---|---|---|---|
| Intercept | 119.901 | 4.607014 | 26.03 | < 0.0001 | 110.85807 | 128.94393 |
| AlwaysUseMask | −0.689229 | 0.058976 | −11.69 | < 0.0001 | −0.80499 | −0.573468 |
| WithoutInsurance | 1.2962509 | 0.182402 | 07.11 | < 0.0001 | 0.9382216 | 1.6542803 |
| Income | −0.114368 | 0.049094 | −2.33 | 0.0201 | −0.210733 | −0.018004 |

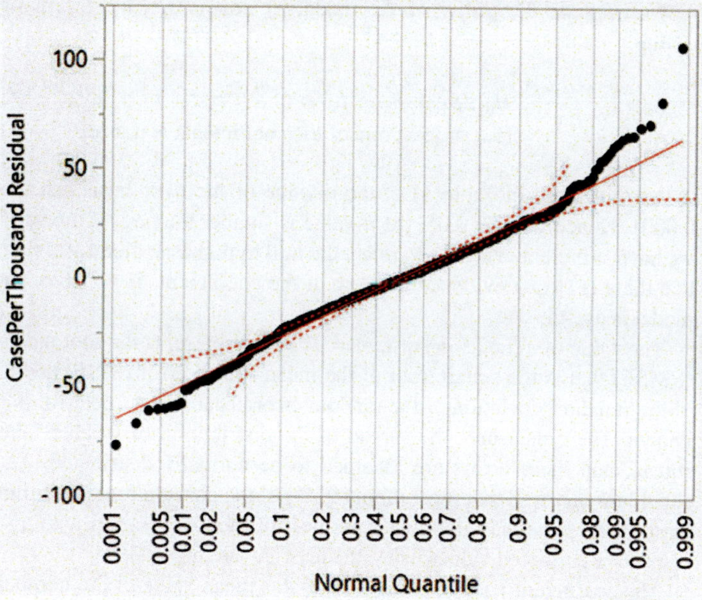


**Fig. 11.7** Residual normal quantile plot

$$\beta_0 = 119.901 \quad \beta_1 = -0.689 \quad \beta_2 = 1.296 \quad \beta_3 = -0.114.$$

Thus, our least-squares prediction is as follows:

$$y = 119.901 - 0.689 \times x_1 + 1.296 \times x_2 - 0.114 \times x_3$$

Figure 11.7 is a graph regarding the probability distribution of the random error component $\varepsilon$. As illustrated in Fig. 11.7, we can see that the residuals do not follow the solid red line exactly, but they are within the Lilliefors confidence bounds and do not contain any pattern [21]. Therefore, the assumption of normality is verified.

Table 11.2 is the analysis of variance showing that $s^2$, the estimator of the variance $\sigma^2$ of the random error term $\varepsilon$, is 455.2. Given the multiple regression model:

$$y = \beta_0 + \beta_1 \times x_1 + \beta_2 \times x_2 + \beta_3 \times x_3$$

**Table 11.2** Analysis of variance

| Source | DF | Sum of squares | Mean square | F Ratio |
|---|---|---|---|---|
| Model | 3 | 125406.55 | 41802.2 | 91.83363 |
| Error | 820 | 373248.97 | 455.2 | **Prob > F** |
| C. Total | 823 | 498555.53 | | <0.0001 |

We evaluate the utility of the model by conducting the following hypothesis testing:

$$H_0 : \beta_1 = \beta_2 = \beta_3 = 0$$
$$\mathrm{H_a} : \text{at least one of the coefficients is nonzero}$$

As illustrated in Table 11.2, the *p*-value of the hypothesis test is smaller than 0.0001. Hence, at $\alpha = 0.05$, the *p*-value is smaller than $\alpha$. As a result, in this case, we have sufficient evidence to reject the null hypothesis; therefore, we can conclude that there is strong evidence that one of the coefficients is nonzero, and the overall model is useful.

The value of *s* (i.e., $s = \text{sqrt}(s^2) = 21.335$) represents the root mean square error (RMSE) that is the square root of the mean square error (MSE) and is an estimate of the standard deviation of the random error because the true error is generally not known. The *s* measures the spread of the distribution of *y*-values about the least-square, and these errors are assumed to be normally distributed. As a result, we should usually find that most (about 95%) of the observations lie within $2 \times s$ of the least-square, that is, $y \pm 42.67$, and most of the *y*-values fall within 42.67 of their respective predicted values using the least-square regression.

The coefficient $\beta_1 = -0.689$ estimates mean value of the number of COVID-19 cases per one thousand in a county (i.e., *E*(*y*)) to decrease 0.689 for each 1% increase of people who always like to wear the mask ($x_1$) when WithoutInsurance ($x_2$) and Income ($x_3$) are held constant. At $\alpha = 0.05$, the *p*-value is less than 0.0001, which is less than $\alpha$. As a result, in this case, we have sufficient evidence to reject the null hypothesis; therefore, we can conclude that the mean value of the number of COVID-19 cases per one thousand in a county (*E*(*y*)) will decrease as mask use ($x_1$) increases while all the other variables are held constant.

The coefficient $\beta_2 = 1.296$ estimates the mean value of the number of COVID-19 cases per one thousand in a county (i.e., *E*(*y*)) to increase 1.296 for each one percent increase of people who do not have health insurance ($x_2$) when AlwaysUseMask ($x_1$) and Income ($x_3$) are held constant. At $\alpha = 0.05$, the *p*-value is less than 0.0001 which is less than $\alpha$. As a result, in this case, we have sufficient evidence to reject the null hypothesis; therefore, we can conclude that the mean value of the number of COVID-19 cases per one thousand in a county (*E*(*y*)) will increase as WithoutInsurance ($x_2$) increases while all the other variables are held constant.

The coefficient $\beta_3 = -0.114$ estimates mean value of the number of COVID-19 cases per one thousand in a county (i.e., *E*(*y*)) to decrease 0.114 for each one

**Table 11.3** Parameter estimates

| Term | Estimate | Std. error | *t* Ratio | Prob > \|*t*\| | Lower 95% | Upper 95% |
|---|---|---|---|---|---|---|
| Intercept | 2.0099036 | 0.140897 | 14.27 | <0.0001 | 1.7333428 | 2.2864644 |
| AlwaysUseMask | 0.0022637 | 0.001804 | 1.26 | 0.2098 | −0.001277 | 0.005804 |
| WithoutInsurance | 0.0171876 | 0.005578 | 3.08 | 0.0021 | 0.006238 | 0.0281373 |
| Income | −0.013951 | 0.001501 | −9.29 | < 0.0001 | −0.016898 | −0.011004 |

thousand-dollar increase ($x_3$) when AlwaysUseMask ($x_1$) and WithoutInsurance ($x_2$) are held constant. At $\alpha = 0.05$, the *p*-value is 0.0201, which is less than $\alpha$. As a result, in this case, we have sufficient evidence to reject the null hypothesis; therefore, we can conclude that the mean value of the number of COVID-19 cases per one thousand in a county ($E(y)$) will decrease as the median household income ($x_3$) increases while all the other variables are held constant.

The intercept value of $\beta_0$ is a marker, and in most cases, unless all the $x's$ are zero, then it is mainly used to set up the value of *y* points. A 95% confidence interval for $\beta_2$ is in the range of 0.938 to 1.654, that is to say, there is 95% chance that our estimation for $\beta_2$ is going to occur in that range, and we have the mean value of the number of COVID-19 cases to increase in anywhere from 0.938 to 1.654 for every one percentage increase of people who have no healthcare insurance coverage ($x_2$) when AlwaysUseMask ($x_1$) and Income ($x_3$) are held constant.

The SAS script about the multiple regression model is attached in the Appendix. To investigate more on the relationship among the number of COVID-19 deaths, healthcare insurance coverage, median household income, and the mask use, we replace the dependent variable into number of COVID-19 deaths and conduct an analysis on the results of the regression model. Table 11.3 gives the parameter estimates regarding the modeling with least-squares where can find that the variable AlwaysUseMask is not significant due to a *p*-value of 0.2098, which is larger than 0.05. This makes sense since wearing mask helps to reduce the spread of COVID-19 diseases; however, once people get infected, it would not help to reduce the mortality rate. The details on the SAS script for this modeling are illustrated in the Appendix.

## 11.6 Conclusions and Future Work

The poorer and more vulnerable social groups have a disproportionately higher number of deaths, implying that an increase of income and health insurance coverage inequality contributes to an increase of COVID-19 incidences. In this paper, we validate this inequality phenomena using data visualization and multiple regression modeling. One of the major contributions of this work is that we create a COVID-19 dataset including median household income, healthcare insurance coverage, and mask-use behaviors by county. This dataset we build is publicly available so people in the healthcare professional domain can use it as a reference to help policymakers make decisions wisely. The other contributions include the following findings: (1) the higher the median household income in a region, the lower the number of COVID-19 cases and deaths in that region; (2) the higher the healthcare insurance coverage in a region, the lower the number of COVID-19 cases and deaths in that region, and (3) the higher the percentage of people who always wear masks, the lower the number of COVID-19 cases in that region, and thus wearing masks does help to reduce the spread of COVID-19 virus.

Wearing mask can reduce the spread of COVID-19, not only extremely important for high-risk groups for COVID but could also be one of the most effective solutions to mitigate immediately the significant implications for this socioeconomic inequality. In the future, we will integrate other features such as unemployment rate, age groups, and education level in a region and then apply multiple regression and other machine learning technologies, such as E-means clustering algorithm [22–24], to build a predictive model to predict the number of patients so healthcare officials and professionals can get notified proactively when making corresponding policies.

**Acknowledgments** Research supported by New Hampshire INBRE through an Institutional Development Award (IDeA), P20GM103506, from the National Institute of General Medical Sciences of the NIH, and in partial by Keene State FDG grant.

## Appendix

The following shows the SAS scripts to generate the multiple regression model among the number of COVID-19 cases, median household income, health insurance coverage, and mask use.

```
Fit Model(
    Y( :CasePerThousand ), Effects( :AlwaysUseMask,
                :WithoutInsurance, :Income ),
    Personality( "Standard Least Squares" ), Emphasis( "Effect
                Leverage" ),
    Run(
        :CasePerThousand << {Summary of Fit( 1 ), Analysis
                of Variance( 1 ),
        Parameter Estimates( 1 ), Lack of Fit( 0 ), Scaled
                Estimates( 0 ),
        Plot Actual by Predicted( 1 ), Plot Regression( 0 ),
        Plot Residual by Predicted( 1 ), Plot Studentized
                Residuals( 0 ),
        Plot Effect Leverage( 1 ), Plot Residual by Normal
                Quantiles( 1 ),
        Box Cox Y Transformation( 0 )}
    ),
    SendToReport(
        Dispatch(
            {"Response CasePerThousand", "Whole Model",
                        "Parameter Estimates"},
```

```
            "Lower 95%", NumberColBox, {Visibility( "Visible" )}
        ),
        Dispatch(
            {"Response CasePerThousand", "Whole Model", "Parameter
                                Estimates"},
              "Upper 95%", NumberColBox, {Visibility( "Visible" )}
        )
    )
)
Fit Model(
    Y( :DeathPerThousand ), Effects( :AlwaysUseMask,
                                :WithoutInsurance, :Income ),
    Personality( "Standard Least Squares" ), Emphasis( "Effect
                                Leverage" ),
    Run(
        :DeathPerThousand << {Summary of Fit( 1 ), Analysis
                                of Variance( 1 ),
        Parameter Estimates( 1 ), Lack of Fit( 0 ), Scaled
                                Estimates( 0 ),
        Plot Actual by Predicted( 1 ), Plot Regression( 0 ),
        Plot Residual by Predicted( 1 ), Plot Studentized
                                Residuals( 0 ),
        Plot Effect Leverage( 1 ), Plot Residual by Normal
                                Quantiles( 0 ),
        Box Cox Y Transformation( 0 )}
    ),
    SendToReport(
        Dispatch(
        {"Response DeathPerThousand", "Whole Model", "Parameter
                                Estimates"},
            "Lower 95%", NumberColBox, {Visibility( "Visible" )}
        ),
        Dispatch(
        {"Response DeathPerThousand", "Whole Model", "Parameter
                                    Estimates"},
            "Upper 95%", NumberColBox, {Visibility( "Visible" )}
        )
    )
)
```

## References


1. COVID-19 Dashboard, https://coronavirus.jhu.edu/map.html, retrieved on April 9, 2022
2. Will the COVID-19 Crisis Exacerbate Inequality? https://www.chicagobooth.edu/review/will-covid-19-crisis-exacerbate-inequality, retrieved on April 9, 2022
3. W. Lu, M. Tavallaee, A.A. Ghorbani, Detecting network anomalies using different wavelet basis functions, in *6th Annual Communication Networks and Services Research Conference (CNSR 2008)*, (2008), pp. 149–156. https://doi.org/10.1109/CNSR.2008.75
4. W. Lu, A novel framework for network intrusion detection using learning techniques, in *PACRIM. 2005 IEEE Pacific Rim Conference on Communications, Computers and Signal Processing, 2005*, (2005), pp. 458–461. https://doi.org/10.1109/PACRIM.2005.1517325

5. W. Lu, I. Traore, A new unsupervised anomaly detection framework for detecting network attacks in real-time, in *Cryptology and Network Security*, ed. by Y.G. Desmedt, H. Wang, Y. Mu, Y. Li, (Springer Nature, 2005), pp. 96–109. isbn:978-3-540-32298-6
6. W. Lu, I. Traore, An unsupervised approach for detecting DDoS attacks based on traffic based metrics, in *Proceedings of IEEE Pacific Rim Conference on Communications, Computers and Signal Processing (PACRIM 2005)*, (Victoria, B.C., August 2005), pp. 462–465
7. D. Hawkins, Social determinants of COVID-19 in Massachusetts, United States: An ecological study. J. Prev. Med. Public Health **53**(4), 220–227 (2020). https://doi.org/10.3961/jpmph.20.256
8. N. Krieger, P.D. Waterman, J.T. Chen, COVID-19 and overall mortality inequities in the surge in death rates by zip code characteristics: Massachusetts, January 1 to May 19, 2020. Am. J. Public Health **110**(12), 1850–1852 (2020). https://doi.org/10.2105/AJPH.2020.305913. Epub 2020 Oct 15. PMID: 33058698; PMCID: PMC7662002
9. T.F. Liao, F. De Maio, Association of social and economic inequality with coronavirus disease 2019 incidence and mortality across US counties. JAMA Netw. Open **4**(1), e2034578 (2021). https://doi.org/10.1001/jamanetworkopen.2020.34578
10. O.Y. Bello-Chavolla, A. González-Díaz, N.E. Antonio-Villa, C.A. Fermín-Martínez, A. Márquez-Salinas, A. Vargas-Vázquez, J.P. Bahena-López, C. García-Peña, C.A. Aguilar-Salinas, L.M. Gutiérrez-Robledo, Unequal impact of structural health determinants and comorbidity on COVID-19 severity and lethality in older Mexican adults: Considerations beyond chronological aging. J. Gerontol. A Biol. Sci. Med. Sci **76**(3), e52–e59 (2021). https://doi.org/10.1093/gerona/glaa163. PMID: 32598450; PMCID: PMC7337730
11. S.J. Kim, W. Bostwick, Social vulnerability and racial inequality in COVID-19 deaths in Chicago. Health Educ. Behav. **47**(4), 509–513 (2020)
12. G.A. Millett, A.T. Jones, D. Benkeser, S. Baral, L. Mercer, C. Beyrer, B. Honermann, E. Lankiewicz, L. Mena, J.S. Crowley, J. Sherwood, P.S. Sullivan, Assessing differential impacts of COVID-19 on black communities. Ann. Epidemiol. **47**, 37–44 (2020). https://doi.org/10.1016/j.annepidem.2020.05.003
13. D.B.G. Tai, A. Shah, C.A. Doubeni, I.G. Sia, M.L. Wieland, The disproportionate impact of COVID-19 on racial and ethnic minorities in the United States. Clin. Infect. Dis **72**(4), 703–706 (2021). https://doi.org/10.1093/cid/ciaa815. PMID: 32562416; PMCID: PMC7337626
14. C. Oronce, C.A. Scannell, I. Kawachi, Y. Tsugawa, Association between state-level income inequality and COVID-19 cases and mortality in the USA. J. Gen. Int. Med. **35**(9), 2791–2793. https://doi.org/10.1007/s11606-020-05971-3
15. Mask-Wearing Survey Data, https://github.com/nytimes/covid-19-data/tree/master/mask-use, retrieved on April 9, 2022
16. COVID-19 Inequality Data, https://unh.box.com/s/uuyfuhhs4r7pjo0h7vwerlynhong4oaj, retrieved on April 11, 2022
17. W. Lu, H.J. Tong, Detecting network anomalies using CUSUM and EM clustering, in *Advances in Computation and Intelligence*, ed. by Z. Li, (Springer Nature, 2009), pp. 297–308. isbn:978-3-642-04843-2
18. W. Lu, I. Traore, Unsupervised anomaly detection using an evolutionary extension of K-means algorithm. Int. J. Inf. Comput. Secur. Inderscience Publisher **2**(2), 107–139 (2008)
19. E.C. Alexopoulos, Introduction to multivariate regression analysis. Hippokratia **14**(Suppl 1), 23–28 (2010). PMID: 21487487; PMCID: PMC3049417
20. A. DeBono, *Research Methods and Statistics for the Social Sciences: A Brief Introduction* (Cognella Academic Publishing, 2020). isbn-13:978-1516537389
21. Multiple Linear Regression, https://www.jmp.com/en_nl/learning-library/topics/correlation-and-regression/multiple-linear-regression.html, retrieved on August 18, 2022
22. W. Lu, H. Tong, I. Traore, E-means: An evolutionary clustering algorithm, in *Advances in Computation and Intelligence. ISICA 2008. Lecture Notes in Computer Science*, ed. by L. Kang, Z. Cai, X. Yan, Y. Liu, vol. 5370, (Springer, Berlin, Heidelberg, 2008), pp. 537–545. https://doi.org/10.1007/978-3-540-92137-0_59

23. W. Lu, I. Traore, Determining the optimal number of clusters using a new evolutionary algorithm, in *Proceedings of IEEE International Conference on Tools with Artificial Intelligence (ICTAI 2005)*, (Hongkong, November 2005), pp. 712–713
24. W. Lu, I. Traore, A new evolutionary algorithm for determining the optimal number of clusters, in *Proceedings of IEEE International Conference on Computational Intelligence for Modeling, Control and Automation (CIMCA 2005)*, vol. 1, (2005), pp. 648–653

# Chapter 12
# Cellular Communication Network Evolution and the Reliability of System Design from 1G to 6G

**Rashed Hasan Ratul and Hwang-Cheng Wang**

## 12.1 Introduction

Prior to the invention of cellular technologies, push-to-talk (PTT) technique was used in early mobile telephone systems back in the 1940s in the United States. In that era, communications were performed over a single channel via a half-duplex mode that could barely support efficient data transmission and reception. Later, a dual channel was introduced, separating transmission and reception channels, allowing full-duplex operation in mobile communication. The forerunner of 1G technology is the mobile radio telephone (i.e., "0G"), where transportable phones would connect to a centralized operator. This pre-1G mobile communication system got its first breakthrough when coverage regions were separated into multiple cells allowing frequency reuse to enhance the overall capacity [1].

Following that, cellular communication technologies have gone through an exponential growth from 1G to 5G over the last 40 years. The local voice service was launched in the 1980s called NMT (Nordic Mobile Telephone) by the very first generation (1G) of cellular technology based on analog systems. However, a digital system was incorporated in the second generation (2G) of cellular technologies

R. H. Ratul
Islamic University of Technology (IUT), Dhaka, Bangladesh
e-mail: rashedhasan@iut-dhaka.edu

H.-C. Wang (✉)
National Ilan University (NIU), Yilan, Taiwan
e-mail: hcwang@niu.edu.tw



I. Woungang, S. K. Dhurandher (eds.), *5th International Conference on Wireless, Intelligent and Distributed Environment for Communication*, Lecture Notes on Data Engineering and Communications Technologies 174,
https://doi.org/10.1007/978-3-031-33242-5_12

called GSM (Global System for Mobile communication) in the 1990s, where voice service and some more features like SMS and MMS were included. Later, GPRS (2.5G) and Enhanced Data rate for GSM Evolution (EDGE) (2.75G) were added as the updated variant of 2G in order to meet higher data rate demand and internet facilities.

Third-generation (3G) networks were launched in the early 2000s with enhanced services in voice calls, SMS, and MMS with an improved bandwidth resulting in a much higher data rate in the internet uses experienced by end users [2]. The later releases of 3G, known as 3.5G and 3.75G, served to upgrade 3G performances in specific domains. Further, an IP-based technology released in the late 2010s with much higher bandwidth and data rate took place in the cellular communication network named 4G (LTE-A) though LTE was first incorporated with 3.9G that was eventually commercialized as 4G in most of the countries.

In upcoming years, fifth-generation (5G) technologies are promising to take place with massive machine type communications (mMTC), enhanced mobile broadband (eMBB), and ultra-reliable and low latency communications (URLLC). These features of 5G technologies would potentially offer versatile applications such as remote control vehicles, automated driving, 3D videos, home automation, and smart cities [1]. Sixth generation is the next-generation advanced cellular communication systems that will function as a scattered neural network, providing communication linkages to connect the physical, cyber, and natural worlds, ushering in an era in which everything will be felt, linked, smart, and smooth. This, in turn, will provide a solid platform for future Intelligence of Everything (IoE).

Sixth-generation technologies would be superfast and far beyond the need and capacity of human demands; instead, they would be implemented with native AI, networked sensing, extreme connectivity, and sustainability to develop IoE infrastructure in the near future. Numerous developed nations have started particular projects to promote 6G. Most countries are targeting to commercialize 6G by 2030. However, despite the fact that advanced technologies like 3G and 4G are now widely deployed worldwide as of 2022, GSM (2G) is still very popular in remote places worldwide in the mobile market.

Undoubtedly, technological developments contribute to improving many aspects of human life, such as faster and more dependable communication. This paper focuses on the reliability aspect of advanced cellular generation technologies, and it is organized as follows. Section 12.2 provides a short overview of all the cellular generations. Then, dependability related to different stages of communication is described in Sects. 12.3 and 12.4. Section 12.5 discusses a numeric calculative approach to highlight network advances. Section 12.6 presents the simulation results and discussion. Finally, Sect. 12.7 concludes this paper.

## 12.2 Overview of the Cellular Generations

### *12.2.1 First Generation (1G)*

First-generation cellular operations employed analog transmission exclusively for voice services. Several alternative 1G cellular standards were deployed in various nations, but the Nordic Mobile Telephone (NMT) and Advanced Mobile Phone System (AMPS) systems were the most broadly adopted worldwide.

A method known as FDMA (Frequency Division Multiple Access) was utilized with the 1G mobile system for voice operations only [2]. It had a very simple infrastructure with inadequate capacity, poor reception, unsatisfactory battery life, high background noise, high interference, and lower SINR. Moreover, it was highly insecure because a third party with an all-band radio receiver could decrypt the signals.

### *12.2.2 Second Generation (2G)*

Second-generation cellular networks were commercially launched on the GSM standard in Finland back in 1991. Second-generation (2G) systems utilize digital multiple access technologies such as TDMA and CDMA compared to first-generation (1G) systems. Consequently, digitally encrypted phone conversation, efficient use of radio frequency spectrum enabling more users per frequency band, and better data service were offered by the 2G system, which signifies higher benefits over their predecessors [2].

GSM was the first 2G network, offering both voice and data services. GPRS, also referred to as 2.5G, is a development in GSM technology. Following 2.5G, EDGE (Enhanced Data rate for GSM Evolution) technology was released under the authority of 2G, commonly known as 2.75G, which was another improvement in GPRS. Circuit-switched networks are utilized in 2G to transmit and receive data from source to destination for voice and data. In 2G, voice quality and interference from noise were also better. Second generation saw the first use of digital encryption for safe data transport. The core network part of 2G and 3G is almost identical, as shown in Fig. 12.1.

### *12.2.3 Third Generation (3G)*

With a view to expediting the network speed, the earliest 3G protocols were WCDMA and UMTS. In 2001, SK Telecom in South Korea made its first commercial release. With three primary frequency bands of 850, 1900, and 2100 MHz, it followed the packet switching paradigm and used CDMA technology. The

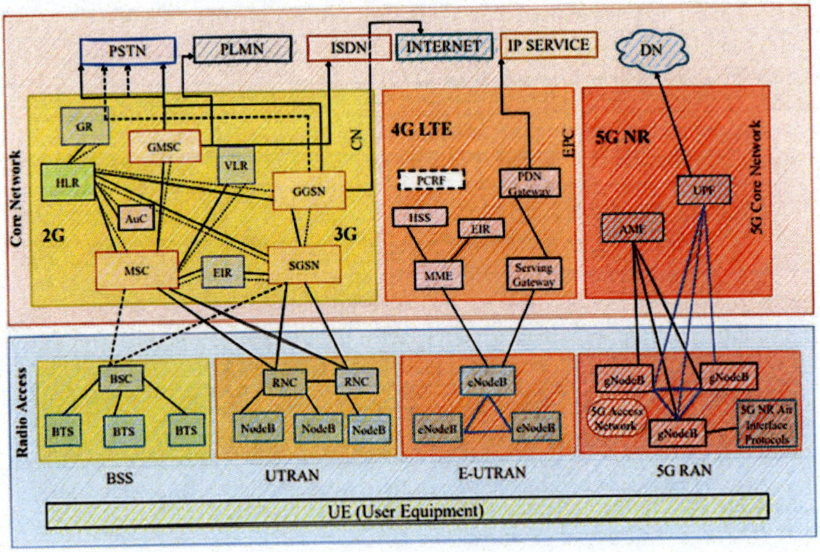


**Fig. 12.1** Combined network architecture of 2G, 3G, 4G, and 5G

WCDMA standard was based on GSM. UMTS was a new standard developed by 3GPP that was primarily utilized throughout Europe [3].

With a noteworthy escalation in transmission speed, 3G facilitated video calling, mobile internet, and streaming facilities. Another significant change with 3G was that individuals did not have to pay for services based on the time period to use the service; instead, they could pay based on the total amount of data usage. Two other standards are known as HSPA or 3.5G and HSPA plus or 3.75G. HSPA was an upgraded form of WCDMA technology. Digital reliability and data security have been improved in 3G communication technology. The advanced version of HSPA, known as HSPA plus, featured multiple input multiple output (MIMO). When a 4G network is unavailable, the symbol for that is now sometimes displayed, which is $H^+$. The RAN (Radio Access Network) part of 3G is slightly modified and updated than 2G, as shown in Fig. 12.1. In 3G, the RNC (Radio Network Controller) remains connected with each other to manage the base stations in an efficient and reliable manner.

### *12.2.4 Fourth Generation (4G)*

The fourth generation of wireless technology is also known as LTE. The first commercially available 4G LTE networks were in Oslo, Norway, and Stockholm, Sweden. But 4G was more than just LTE; it also contained another standard known as WiMAX, established by IEEE and was known as IEEE 802.16 when it was installed in South Korea in June 2006 [4].

LTE is a fully packet switched network that uses orthogonal frequency division multiple access (OFDMA) and multiple input multiple output (MIMO) [1]. LTE offers a wide range of use cases, including IP telephone and voice over IP (VoIP), 3D television, video conferencing with numerous active participants, mobile web connection that is simple and quick, HD mobile television, online gaming, and even cloud computing [5]. The network architecture of 4G has gone through extensive changes and modifications. The base stations here are interconnected with each other, and the RNC or BSC is no longer available in the RAN part, as shown in Fig. 12.1. Moreover, 4G service can switch back to 3G or 2G services according to the radio link quality available in a specific environment. As a result, instead of network failure in a poor radio link environment, LTE offers an adaptive network service with respect to the overall cellular network availability. Overall, its adaptability ensures a much more reliable performance over the previous cellular generations.

### *12.2.5 Fifth Generation (5G)*

From the very beginning of cellular technology, the later versions were developed to mitigate the shortcomings of the previous generations. However, 5G is not an upgraded version of 4G; rather, it is an entirely new technology. People once believed that mobile telecommunication networks now could meet all of their needs, but when new technology ideas emerged, such as the Internet of Things (IoT), driverless cars, smart cities, and a large number of sensors, the demand for a more advanced wireless technology rapidly showed up. Fifth generation is regarded as the solution for all dimension applications, which cellular phone companies began deploying worldwide in 2019 [6].

The benefits of 5G include its adaptability to any special purpose network or device, its aptitude to function as a distributed data center for processing, its ability to be customized for each use case, its capability to offer prompt response, and so on. The simplified network architecture of 5G is shown in Fig. 12.1.

### *12.2.6 Sixth Generation (6G)*

While the deployment of next-generation mobile communication systems is still 10 or more years away, many existing 6G programs and accompanying investments offer an enticing promise for significant acceleration of 6G studies, which will approximately go live in 2030. Some emerging applications like terahertz communications, ubiquitous coverage (terrestrial, air, interplanetary, and marine), tactile/haptic communications, national security, imaging and sensing, and cyber-physical systems are the sectors that 6G will be expected to thrive [7]. Sixth

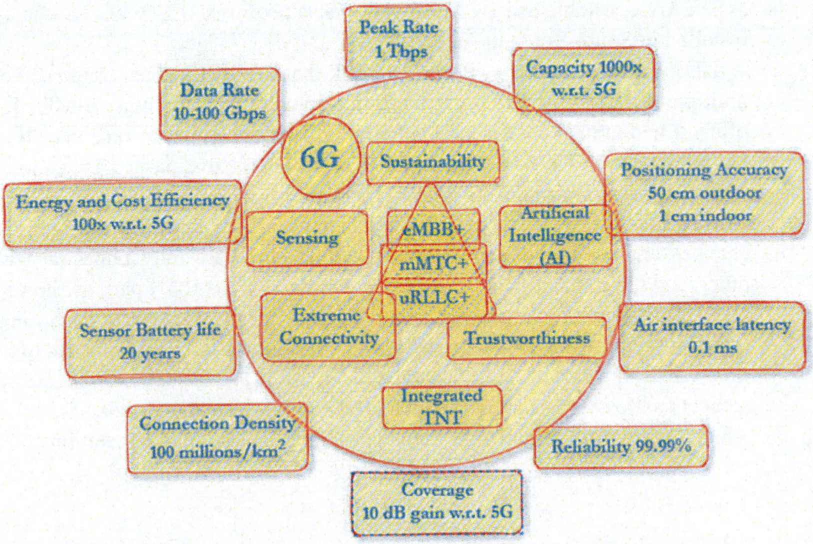


**Fig. 12.2** Expected features of 6G technologies

**Table 12.1** Generalized frequency range of different cellular generations [8]

| Generation | Frequency |
|---|---|
| 1G 2G 3G 4G | Frequency up to 6 GHz (Sub 6 GHz) |
| 5G | 410–7125 MHz (FR1) |
| | 24,250–52,600 MHz (FR2-1) |
| | 52,600–71,000 MHz (FR2-2) |
| 6G | Frequency above 100 GHz, even up to 3 THz |

generation will incorporate sensing and AI that will expand the capabilities substantially (Fig. 12.2). Frequencies up to 6 GHz are considered for 2G, 3G, and 4G. On the other hand, 5G is supported by various ranges of frequencies (Table 12.1) as per 3GPP release-18. Frequencies up to 100 GHz are considered for 5G. Frequencies above 100 GHz, even up to 3 THz, are considered for 6G [9].

## 12.3 Dependability Paradigms

### *12.3.1 Cell Selection, Reselection, and Handover*

When any UE is first powered up, it has to camp on a cell considering any nearby cell as its serving cell. The UE looks for the most suitable cell inside the chosen Public Land Mobile Network (PLMN) to set up camp whenever a PLMN has been chosen. Using data stored in the UE or applying preliminary cell selection are two methods for performing this cell selection. The UE may initiate quick cell evaluation or cell reselection while camped on a cell of the selected PLMN. If no suitable cell is located, cell selection using stored information is initiated. When a UE exits idle mode to enter connected mode, it does an immediate cell evaluation to determine the best cell for the access attempt. On return to idle mode, the UE chooses to camp on the best cell of the cells used when leaving connected mode [10]. As the UE changes its position, the serving cell gets changed according to the mobility. Two types can be considered here while changing a cell, as shown in Table 12.2.

The decision of the change of cell depends on the current radio link condition experienced by a UE from the serving cell compared to the neighboring cells. As radio signals are often time-varying, relative signal strength between two cells can be transient fluctuation. Thus, the decision for cell switching must be considered when the neighboring target cell provides better quality than the current serving cell for a sufficient margin continuously for a sufficient amount of time period. Otherwise, there is a good possibility of switching back and forth between cells, known as the ping-pong effect. The ping-pong effect can reduce the overall system performance by using resources unnecessarily. To reduce the ping-pong effect, when a neighboring cell gets better by the margin of $Q_{\mathrm{Hyst}}$, then consider cell switching. Second, a timer called $T_{\mathrm{reselection}}$ is also incorporated in order to assure efficient and reliable data transfer with good radio link quality.

### 12.3.2 Cell Selection Criteria (S Criterion)

Cell Reselection/Handover can be classified into three main categories such as:

- Cell Reselection/Handover for intra-LTE frequencies
- Cell Reselection/Handover for inter-LTE frequencies
- Cell Reselection/Handover for inter-RAT frequencies

The cell selection value, $S$, is

$$S = Q - Q_{\min} - P_{\mathrm{compensation}} \tag{12.1}$$

The cell selection criterion is fulfilled if $S > 0$ [10]. For any cell to be suitable, $Q_{\mathrm{rxlevmin}}$ has to be greater than a minimum level, as shown in Table 12.3. According to 3GPP 36.331, LTE cell reselection uses priority-based levels, with Priority 0 being the lowest and Priority 7 being the highest. SIB3 (System

**Table 12.2** Cell reselection versus handover

| RRC State | Procedure | Measurement | Decision |
|---|---|---|---|
| RRC_Idle | Cell reselection | UE | UE |
| RRC_Connected | Handover | UE | Base station |

**Table 12.3** Cell selection criteria

| Symbol | Description |
|---|---|
| $S$ | Cell selection value (dB) |
| $Q$ | Quality level of the received signal (dBm) |
| $Q_{min}$ | Minimum prerequisite quality level in the cell (dBm) |
| $P_{compensation}$ | max(MS_Tx_Pwr_Max – $P$_Max, 0) (dB) |
| MS_Tx_Pwr_Max | Maximum Tx power level a UE may use when accessing the cell (delivered in SIB/system information block) (dBm) |
| $P$_Max | Maximum RF output power of the UE (dBm) |
| $Q_{rxlevmin}$ | Minimum prerequisite Rx level in the cell (dBm) |

Information Block) displays this value as the cellReselectionPriority Information Element (IE). The IE value can be found in SIB1. The Priority here is decided based on the channel bandwidth. As a result, the highest channel bandwidth is given the highest reselection priority. Usually, $Q_{rxlevmin}$ = (IE value × 2) [dBm], where $Q_{rxlevmin}$ indicates for cell selection/reselection the required minimum received RSRP (Reference Signal Received Power) level in the cell, and the value of $Q_{rxlevmin}$ can be found in SIB1 and SIB3 [10].

### 12.3.3 Intrafrequency and Interfrequency Cell Reselection Process

Both intrafrequency cells and interfrequency cells with equal priorities are subject to the ranking criterion. If any neighboring signal strength $R_n$ gets higher than the strength of current serving cell $R_s$, UE does not immediately switch to the neighboring cell. Rather, the neighboring cell signal level or signal quality needs to be higher than $Q_{Hyst}$ to initiate $T_{reselectionEUTRA}$ (Fig. 12.3). UE shall only reselect another cell if $R_n > R_s$ during a certain time interval $T_{reselectionEUTRA}$ [11]. After the end period of the timer, cell switching is performed by the UE. Further details are shown in Table 12.4.

$$R_s = Q_{meas} + Q_{Hyst} \tag{12.2}$$

$$R_n = Q_{meas} - Q_{offset} \tag{12.3}$$

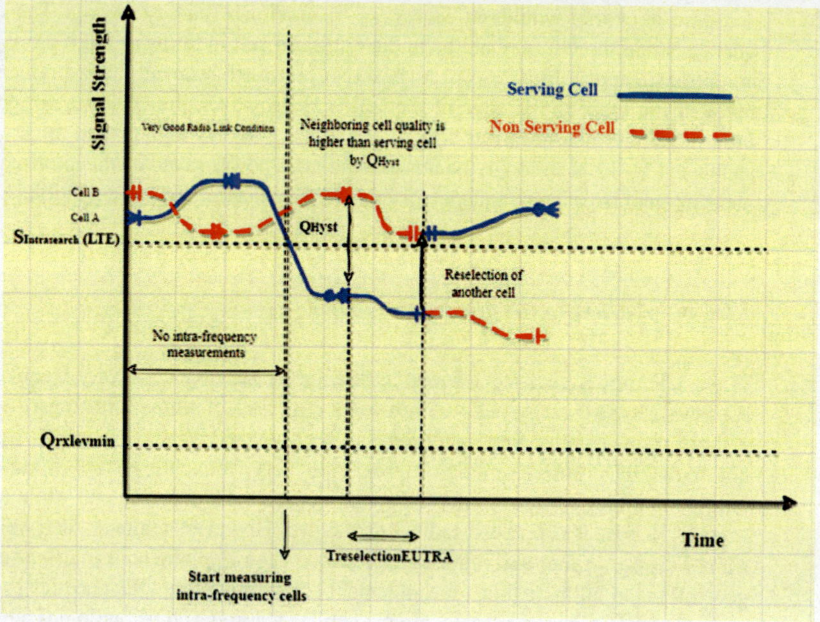


Fig. 12.3 Cell reselection process

**Table 12.4** Explanation of terms in cell reselection process

| Symbol | Description |
|---|---|
| $R_s$ | Serving cell strength |
| $R_n$ | Neighboring cell strength |
| $Q_{meas}$ | RSRP measurement quantity used in cell reselection |
| $Q_{Hyst}$ | A range of constant numbers depending on certain radio link condition |
| $Q_{offset}$ | A value to maintain cell-specific biasing condition |
| $T_{reselectionEUTRA}$ | A specific time period configured by the system |

## 12.3.4 *Measurement Rules and Power Savings During Cell Reselection*

Cell reselection would be initiated only when the current signal level of the serving cell $S$ is lesser than a system configured signal level called $S_{Intrasearch}$. Even if a neighboring cell is far better than the current serving cell, initiation of cell reselection will not take place unless $S < S_{Intrasearch}$.

Some networks would set $S_{\text{Intrasearch}}$ to maximize battery saving. The consequence is that the UE may not be in the optimum cell in terms of radio propagation conditions, especially if $S_{\text{Intrasearch}}$ is set so that intrafrequency measurements are initiated too close to the edge of the cell [12]. However, the UE does not perform intrafrequency measurements if $S$ of the serving cell is higher than $S_{\text{Intrasearch}}$ as shown in Fig. 12.3, although the fact that the extra power saved by the intrafrequency measurement over interfrequency or inter-RAT measurements is negligible.

### 12.3.5 *Adjustments for Speedy UEs*

When UE moves so fast, it needs to change its serving cells very frequently as it passes through many cells within a short period of time. With regard to this, a sharp drop in radio link condition can be observed as $Q_{\text{Hyst}}$ and $T_{\text{reselection}}$ do not allow UE to switch among different cells very frequently just to minimize the ping-pong effect. Moreover, the UE sticks with the older serving cell while moving so fast. Hence, it can cause radio link failure. However, features such as a very short $T_{\text{reselection}}$ timer and reduced $Q_{\text{Hyst}}$ have been implemented in newer cellular generations to mitigate this issue and adapt according to the UE speed [11].

In the RRC_Connected state, almost similar features and technologies have been implemented to ensure reliable connection during the handover procedure. Here, UE sends measurement reports (PCI, RSRP, RSRQ for LTE) to the network, and the network decides everything. UE, however, does not send the measurement report right away. It takes some time to initiate a transfer called TimeToTrigger. This TimeToTrigger can be shortened by scaling it with an appropriate factor (0.25, 0.5) according to the speed of UE to maintain a dependable and trustworthy network connection.

## 12.4 Strategies for Reliability and Network Upgradation

### 12.4.1 *Increased Reliability in Handover Procedure*

In order to guarantee that the user receives the optimum QoS, handover is a sophisticated technique that has to be carefully considered and understood. Generally speaking, handover offers the following:

- Continuity of calls
- Most favorable radio link selection
- Traffic distribution

The UE and RNS (Radio Network Subsystem) will both be used to implement the whole handover process. The UE must undertake measurements of serving radio connection downlink performance and the adjacent cell's received signal intensities and quality. The RNS will receive a signal about these measurements for evaluation. For the UE being served, the RNS measures uplink performance. The RNC (Radio Network Controller) decides on a handover by using measurements along with predetermined thresholds and a handover plan (for UMTS) [13].

### 12.4.2 Handover Strategy for Improved Reliability in 3G, 4G, and 5G

UE connects to RAN/UTRAN/E-UTRAN/CRAN wirelessly. In 2G and 3G, multiple Base Stations (BTS for 2G and NodeB for 3G) are managed by a Base Station Controller (BSC) and Radio Network Controller (RNC), respectively. However, in 4G and 5G, there are no secondary controllers (BSC or RNC) for base stations (eNodeB/gNodeB). Rather, the base stations remain interconnected in the Radio Access Network (RAN) part (Fig. 12.4). The interconnection among the base stations assists in avoiding packet loss while performing Hard Handover. But, in 2G, while performing Hard Handover procedure, some data packet loss is unavoidable as the connection with the new base station is made after breaking the initial connection with the earlier one (Break before Make-Hard Handover).

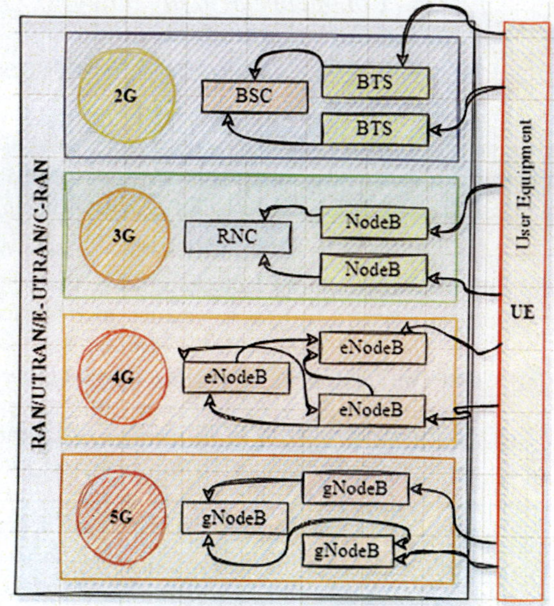


**Fig. 12.4** Comparison of network architectures

To mitigate this packet loss issue, Soft Handover was introduced in 3G where connection with a new base station is to be made before breaking the connection with the older one (Make before Break-Soft Handover). During the period of Soft Handover procedure, there had to be a specific timeframe where the UE is connected with both/multiple base stations. Consequently, the whole process required higher data rates as multiple data packets needed to be sent to both base stations simultaneously. Moreover, it turned out to be highly expensive. As a result, Hard Handover Procedure was reintroduced into 4G technology with the solution of packet loss.

As the base stations remain interconnected in 4G, the initial base station stores the data packets (Downlink) before getting disconnected with the UE. After switching to a new base station, the old one transfers the stored data packets to the new base station as they have mesh connections among themselves. Hence, with slight modification in the network architecture (i.e., interconnection among the base stations), 4G and 5G mitigate the packet loss issue in Hard Handover [11]. The network's handover strategy for the radio link controller regulates the handover decision, which is based on the measurement reports supplied by the UE/base station and other parameters established for each cell. Other than radio link control, network-directed handover may occur to govern traffic distribution across cells. The specific handover tactics are regulated by the cellular network operators.

### 12.4.3 *Commencement of Handover Process and Requirements*

The following is a non-exhaustive list of causes for the initiation of a handover process [13]:

- Channel quality of uplink (UL), signal level of uplink (UL)
- Channel quality of downlink (DL), signal level of downlink (DL)
- Relative distance between UE and base station, change of service
- Suitable cell, directed retry, traffic congestion, preemption

The successful operation of a network depends on the reliability of handover in all forms. The general requirements for executing handover preparation and execution shall include the following: speedy detection of neighboring cells, firm synchronization to neighboring cells, reporting of an adequate number of neighboring cells, instant detection of weak radio link quality level, effective measurement protocols of the serving cell and target neighboring cells, dependable and speedy response mechanisms, efficient synchronization mechanism, speedy and safe discharge of resources, safeguards for unsuccessful handover process, negligible interruption to service, marginal downfall to radio link quality, minimal service drops to other consumers, and comprehensive flexibility to flawlessly handle all the spectrum assigned to a specific network [13].

### *12.4.4 Different Procedures to Enhance Packet Loss Mitigation*

The UE must implement a process to track the receiving end signal strength and quality in downlink (DL) from its serving cell and the downlink signal level and identification of the handover neighboring cells. Different kinds of sophisticated procedures, such as scanning algorithms, heterogeneous handover algorithms, vertical handoff decision algorithms, and so on, have been suggested to mitigate packet losses and obtain higher reliability during the handover procedure [14–16]. Higher generation mobile communication technologies can adopt thesc proposed algorithms in a suitable manner. Furthermore, the adaptability is much higher in upgraded generations compared to the older versions of cellular networks. Hence, the dependability gets much higher in updated and recent cellular generations such as 4G and 5G.

## 12.5 Modifications, Calculations, and Multiple Access Techniques

### *12.5.1 Beamforming*

Fifth generation introduced an access network option called CU/DU (Centralized Unit/Distributed Unit) that can be considered almost equivalent to an RNC. Different protocols are split into CU/DU. A gNodeB can have one CU, and under that CU, there can be 100 or more DUs (Fig. 12.5). Usually, many cells and beams are accumulated within a single DU. In this way, different beams can be used in different directions of the antennas, resulting in beamforming that ensures frequency reuse. With the help of beamforming, wider radiation patterns can be avoided, resulting in lower power level and interference, higher antenna gain, and higher spectral efficiency, so for a narrow radiation pattern, higher CINR can be obtained [11].

### *12.5.2 New RRC State*

Fifth generation also introduced a new RRC state called RRC-inactive. In the previous generations, only RRC-idle and RRC-connected states were available in order to achieve battery savings and data transfer resumption, respectively. RRC-inactive state in 5G offers an option between data transfer recommencement and battery savings that ensures a higher data rate with much healthier battery life. Hence, much better reliability in terms of battery life can be achieved.

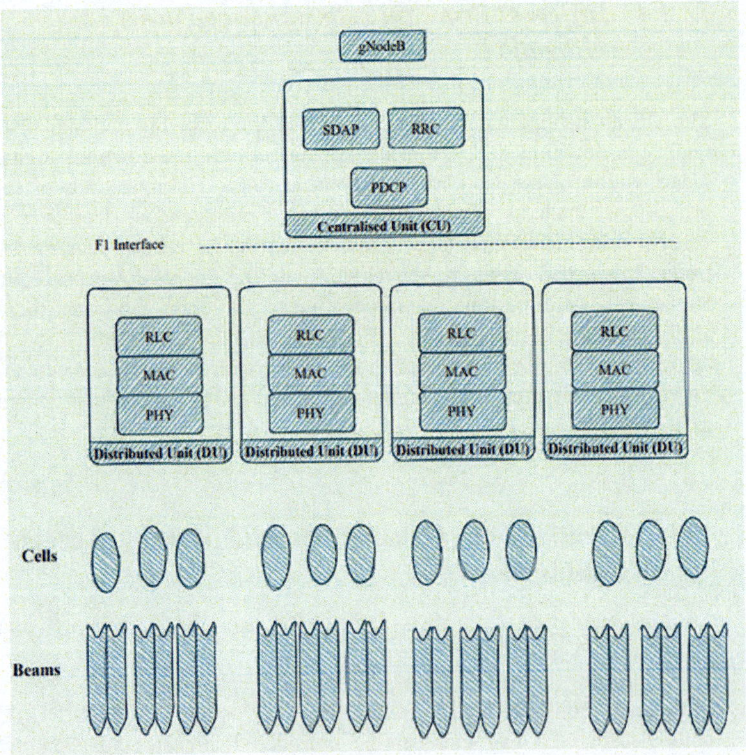


**Fig. 12.5** Extra features of 5G network architecture

### *12.5.3 Higher Spectral Efficiency (from FDMA to CDMA)*

First generation was implemented with FDMA, whereas 2G (GSM) incorporated both FDMA and TDMA multiplexing schemes. With a rapid increase in user demands, earlier multiplexing techniques could not perform well enough to meet customer requirements in terms of the total number of users in a cell, power control, and spectral efficiency [17].

Let us consider a situation where a cellular operator has bought/leased a 15 MHz spectrum from the government. Now, let us analyze how efficiently the occupied spectrum can be handled by different multiple access techniques. Consider FDMA (for 2G GSM), CDMA (for 3G), and OFDMA (for 4G LTE and later generations) [18].

In FDMA, all frequency bands are divided into 200-KHz bandwidth channels that can be time shared by eight people at best. Given that we have a 15 MHz chunk, 75 separate channels of 200 KHz will be available here.

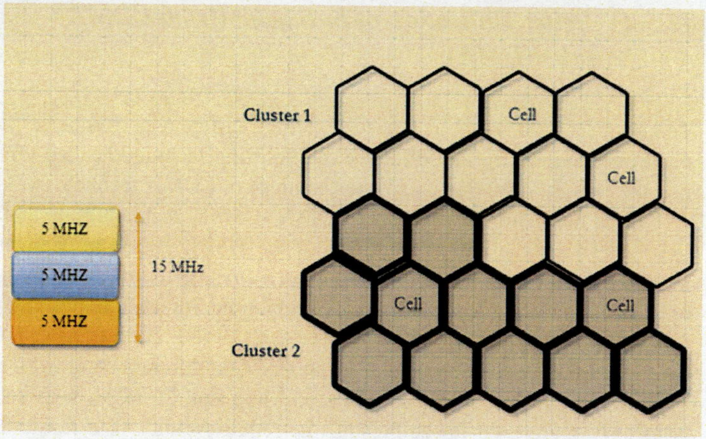


**Fig. 12.6** Cluster for FDMA and frequency chunks for CDMA

$$\left\{\text{Total number of 200-KHz channels } M = \left\{15 \text{ MHz} \div 200 \text{ KHz}\right) = 75\right\}$$

In order to cover a certain geographical area, assume a cluster that can be regarded as the aggregation of the cells over which all segments of the spectrum are used only once. The clusters are repeated one after another to cover a geographical zone (Fig. 12.6).

Let us consider a cluster size $N = 12$ (12 cells in a cluster), so the number of channels available per cell $= M/N = 75/12 = 6.25 \sim 7$. Scaling up to the nearest digit, the number of 200-KHz channels per cell ~7. Hence, total users supported per cell after excluding two channels for control purpose $= (7 \times 8 = 56, 56 - 2 = 54$ users), so only 54 users can be supported per cell via FDMA in 2G.

Now, consider CDMA technology incorporated with 3G mobile networks. In CDMA, only 5 MHz and its integer multiple frequency chunks can be used, and each of the 5-MHz chunks can support approximately 95–100 users. Hence, from the 15 MHz spectrum, three separate chunks of 5 MHz would be acquired that overlap each other, as shown in Fig. 12.6. As a result, almost 285–300 users can be supported per cell via CDMA.

Therefore, the overall spectral efficiency has increased around six times here in terms of the total number of users per cell. Moreover, in CDMA, power control is much better as the base station controls it and sends notifications to a UE 1500 times/s to adjust its transmitting power according to the requirements.

### 12.5.4 Better Utilization of Spectrum (from CDMA to OFDMA)

Though CDMA was sophisticated regarding different performance parameters, a much more advanced and effective multiplexing technique called OFDMA came into the picture, resolving many shortcomings of CDMA technology [17].

One of the main drawbacks of CDMA (UMTS) technology is that it can only use 5 MHz and the integer multiples of the 5 MHz spectrum. Conversely, if any mobile operator purchases, for instance, 18 MHz spectrum from the government, only 15 MHz spectrum can be efficiently used for user allocation purposes in CDMA. The remaining 3 MHz cannot be used for user allocation at all. On the other hand, with OFDMA technology, LTE supports six different bandwidth sizes such as 1.4 MHz, 3 MHz, 5 MHz, 10 MHz, 15 MHz, and 20 MHz, with an appropriate scalability preference depending on the requirements. As a result, in OFDMA, 18 MHz can be fully utilized by splitting it into six separate 3-MHz chunks, which was not possible for CDMA. Hence, the efficiency, performance, and dependability of OFDMA are much higher.

### 12.5.5 Superior Performance and Adaptability from CDMA to OFDMA

CDMA uses the same frequency at the same time for all different communications. Here, every connected pair in a call is assigned a particular code. The codes are orthogonal to each other (i.e., completely different codes from each other). These orthogonal codes are assigned in each cell and are prepared either from OVSF (Orthogonal Variable Spreading Factor) codes (3gpp) or Walsh code (3gpp2) [11].

Another drawback of CDMA is that while transmitting bits/data from one end to another, the data is multiplied/spread with those assigned codes. In the case of transmitting $X$-bit data, another $Y$-bit code needs to be transmitted along with each bit of the original data [17]. Therefore, for an $X$-bit data transfer, the total number $Z$ of bits that need to be transmitted is $XY$.

At the receiver, a composite signal is received that needs to be reprocessed further to retrieve the original data. This process of multiplication of the orthogonal code is called spreading, and the multiplied codes are associated with specific spreading factors (SF). However, SF is decided based on data rate demand, radio link quality, and fairness of radio resource allocation. High SF can accommodate more users in a cell, but the overall data rate would be lower, whereas lower SF would account for the exact opposite scenario. Hence, there is always a tradeoff between data rate and the number of users that can be supported in a cell in CDMA.

On the other hand, OFDMA uses multiple subcarriers for the same users and achieves flat fading. The center frequency experiences zero interference from other subcarriers. In this multiplexing technique, the overall performance is much superior

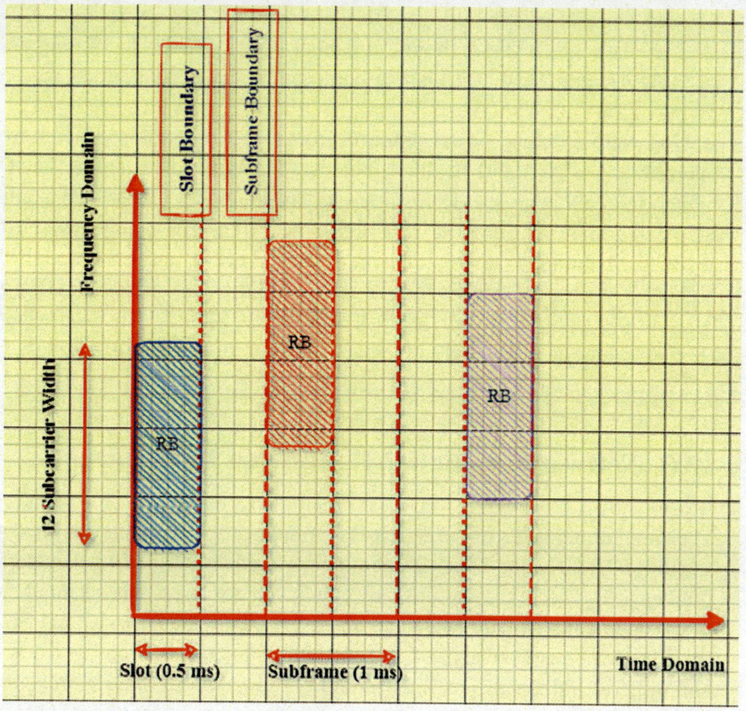


**Fig. 12.7** Resource block allocation strategy

in terms of adaptability. In OFDMA, the network decides the scheduling of resource blocks (RB) at every sub-frame boundary (1 ms) to ensure appropriate data rate and QoS. But it allocates the RB at every slot level (0.5 ms), and different RBs are allocated to different users in separate slots, as shown in Fig. 12.7.

Thus, by employing OFDMA in LTE, UE can have a shorter TTI (Transmission Time Interval) of 1 ms compared to 10–80 ms TTI in 3G. Shorter TTI is supportive because RB scheduling options are time-varying and depend on data rate demand, propagation environment, and radio link quality [19]. Hence, shorter TTI stretches the option to update more frequently so that UE can adapt well with respect to its surroundings and significantly improve efficiency.

### *12.5.6 Control over Uplink Power Transmission in OFDMA*

Uplink power efficiency is vital as phone batteries might drain out faster if UL power consumption is too much. PAPR (Peak to average power ratio) should be low in UL. However, in DL, the base station is always connected to a power source. Therefore, power consumption in DL is not a big issue. Low PAPR can provide better battery life and efficient power savings. For the purpose of battery savings, SC-FDMA is used in place of OFDMA though it cannot offer as much flat fading as OFDMA, so OFDMA and SC-FDMA are a compromise between performance and power savings [20].

However, efficient power control from the UE end is crucial in terms of interference, SINR, and health concerns. Transmit power from UE is generally higher for cell edge users, resulting in much higher battery consumption, low SINR, and high interference. This high-power transmission can cause some health problems to humans as the transmit power directly transfers through the human body [21]. As the UE moves further away from the base station, the risk gets higher because the transmit power level gets higher. OFDMA can have a better power control technique compared to the previous multiple access techniques.

## 12.6 Software Simulation Results and Discussion

The simulation results of the wireless communication channels are presented in this section at 2.1 GHz and 24.25 GHz frequency signals considering 4G and 5G cellular frequencies. These simulations are conducted to assess the channel behavior in urban microcells with 20 MHz and 800 MHz bandwidths for 4G and 5G frequency environments, respectively. During the simulations, actual outdoor conditions and atmospheric influences, such as humidity, temperature, barometric pressure, and polarization, are taken into account, as shown in Table 12.5. These simulation findings were acquired using the mm-Wave channel model simulator NYUSIM 3.0 [22].

In order to accurately model the wireless communication channel, path loss (PL) is a crucial metric that must be taken into consideration. It is helpful in calculating the link budget and outage. The following equation describes the loss (PL) in a free space path with a reference distance of 1 m and attenuations due to atmospheric limitations [23, 24]:

$$\mathrm{PL}\,[\mathrm{dB}] = \mathrm{FSPL}\ [\mathrm{dB}] + 10n\log_{10}(d) + \gamma\ [\mathrm{dB}] + X_\sigma \tag{12.4}$$

In the above equation, $d$ is the three-dimensional T-R separation distance, $\gamma$ is the atmospheric attenuation, $n$ is the path loss exponent, and $X_\sigma$ is the Gaussian random variable with a standard deviation $\sigma$ in decibel. FSPL represents the free space path loss in dB at a T-R separation distance of 1 m at frequency $f$. According to the aforementioned equation, this model is affected by distance and carrier frequency. Additionally, free space path loss FSPL and attenuation $\gamma$ are described as follows:

$$\mathrm{FSPL}\ [\mathrm{dB}] = 20\log_{10}\left(\frac{4\pi f \times 10^9}{c}\right) \tag{12.5}$$

**Table 12.5** Input parameters for simulation

| Input parameters (channel and antenna) | Values/Status |
|---|---|
| Frequency | 2.1 GHz and 24.25 GHz |
| Channel bandwidth | 20 MHz and 800 MHz |
| Scenario | Urban Microcell |
| Distance range of Tx and Rx | 10–500 m |
| Environment | NLOS |
| Transmit power | 30 dBm |
| Biometric pressure | 1013.25 mbar |
| Humidity | 50% |
| Temperature | 27 °C |
| Polarization effect | Co-polarization |
| Rain rate | 37 mm/h |
| Foliage loss | No |
| Foliage attenuation | 0.4 dB/m |
| Tx and Rx antenna array type | Uniform Linear Array (ULA) |
| Tx and Rx antenna azimuth HPBW | 10° |
| Number of Tx and Rx antennas | 1 |
| Correlation distance | 10 m |
| User track type | Linear |
| Human blockage parameters | N/A |
| Base station height | 32 m |

$$\gamma\,[\mathrm{dB}] = \alpha\,[\mathrm{dB/m}] \times d\ \ [\mathrm{m}] \tag{12.6}$$

In the preceding equations, $c$ represents the speed of light, and $f$ indicates the carrier frequency in GHz, while $\alpha$ is the attenuation factor in decibels per meter for the frequency range of 1–100 GHz.

According to the simulation results, path loss and multipath propagation effects are more prominent at higher frequencies than they are at lower frequencies. Additionally, the UE end received power strength is significantly lower at the higher frequency, as shown in Figs. 12.8, 12.9, 12.10, and 12.11. This is a crucial technical issue that needs to be resolved in order to implement reliable communication at higher frequencies. Hence, utilizing higher antenna array gain and reducing propagation distance can be used to overcome the drawbacks at high-frequency bands.

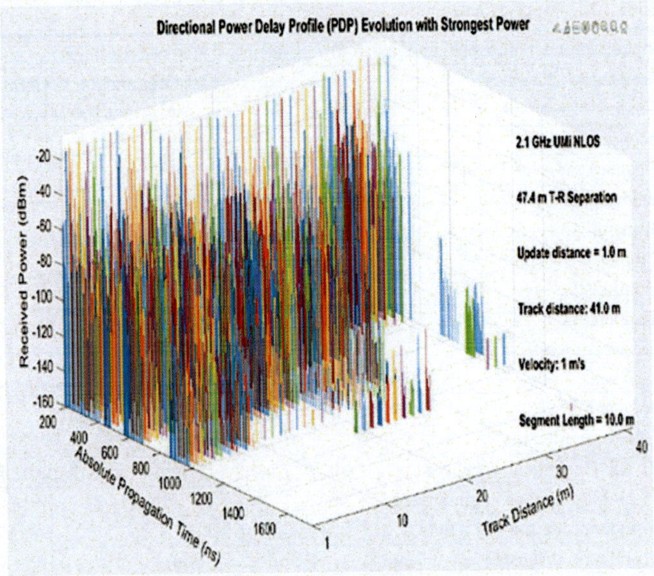


Fig. 12.8 Directional power delay profile (PDP) at 2.1 GHz

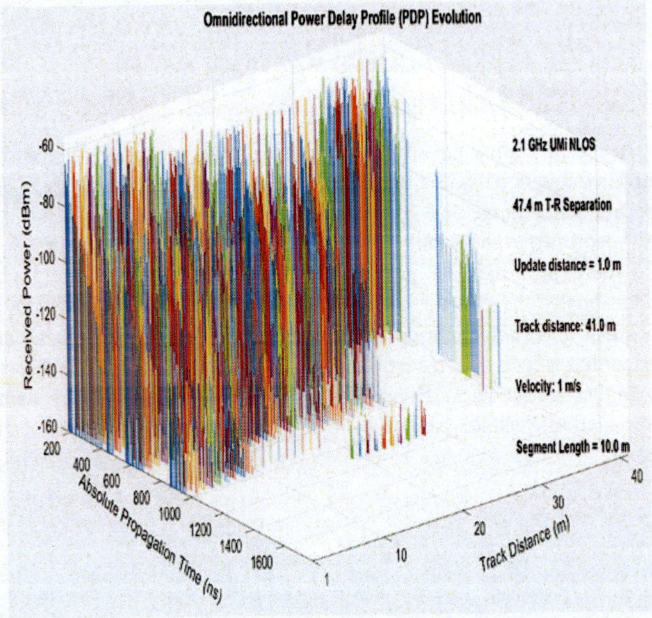


Fig. 12.9 Omnidirectional power delay profile (PDP) at 2.1 GHz

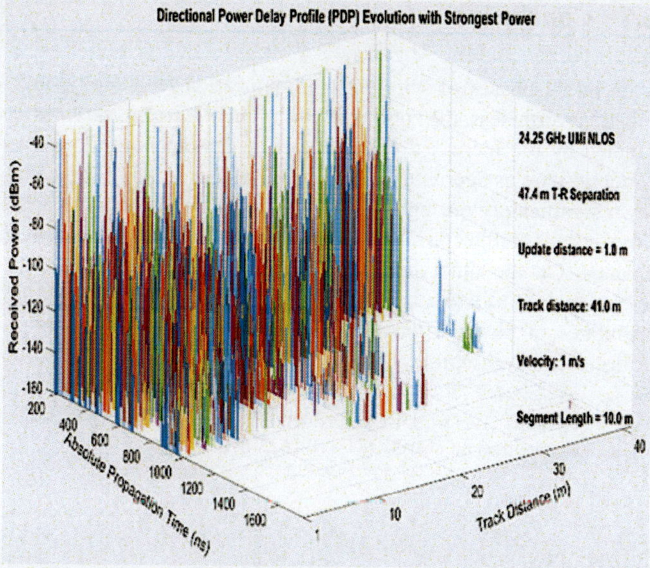


**Fig. 12.10** Directional power delay profile (PDP) at 24.25 GHz

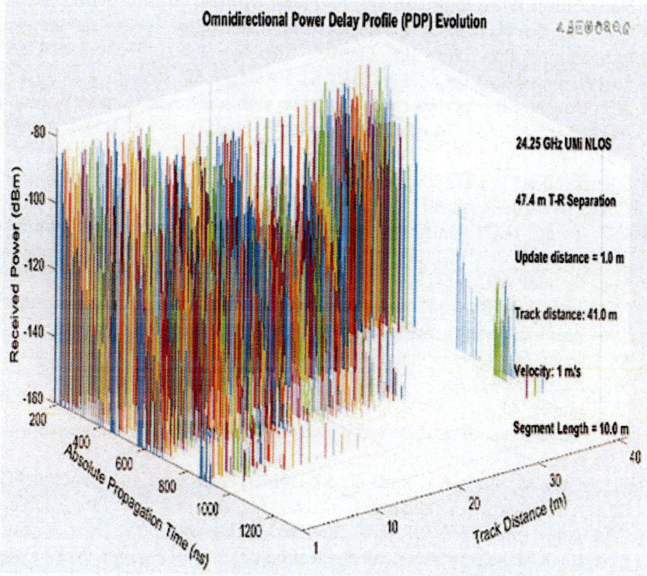


**Fig. 12.11** Omnidirectional power delay profile (PDP) at 24.25 GHz

## 12.7 Conclusion

In the initial phases of wireless communication technology, user demands were limited; nevertheless, the continued evolution of wireless services and infrastructure is fueled by the increased expectations of customers/users. As a result, the field of mobile wireless communication has evolved rapidly. The cellular industry has grown significantly during the past few years by maintaining appropriate stability in the system architecture and multiple access techniques, along with the updated demands. On the other hand, due to the rapid increase of data rate demands, the tendency to utilize a higher-frequency spectrum has caused lesser penetration capabilities of physical barriers. However, the overall reliability, efficiency, and performance have improved enormously over the last decade.

**Acknowledgments** This work was supported in part by the Ministry of Science and Technology, Taiwan, under contract number MOST 110-2622-E-197-001.

## References

1. I.A.M. Balapuwaduge, F.Y. Li, Cellular networks: An evolution from 1G to 4G. Encycl. Wirel. Netw (July), 170–175 (2020). https://doi.org/10.1007/978-3-319-78262-1_214
2. Q.K. Ud Din Arshad, A.U. Kashif, I.M. Quershi, A review on the evolution of cellular technologies. Proc. 16th Int. Bhurban Conf. Appl. Sci. Technol. IBCAST 2019, 989–993 (2019). https://doi.org/10.1109/IBCAST.2019.8667173
3. A.U. Gawas, An overview on evolution of mobile wireless communication networks: 1G-6G. Int. J. Recent Innov. Trends Comput. Commun. **3**(5), 1–5 (2015). [Online]. Available: http://www.ijritcc.org
4. S. Parkvall et al., LTE-advanced – Evolving LTE towards IMT-advanced. IEEE Veh. Technol. Conf., 1–5 (2008). https://doi.org/10.1109/VETECF.2008.313
5. C.C. Tseng, H.C. Wang, F.C. Kuo, K.C. Ting, H.H. Chen, G.Y. Chen, Delay and power consumption in LTE/LTE-A DRX mechanism with mixed short and long cycles. IEEE Trans. Veh. Technol. **65**(3), 1721–1734 (2016). https://doi.org/10.1109/TVT.2015.2405560
6. N. Bairagi, S.K. Bhuyan, A review on adaptive sportswear. Int. J. Res. Publ. Rev. **2**(12), 1053–1064 (2021). [Online]. Available: www.ijrpr.com
7. D. Soldani, 6G fundamentals: Vision and enabling technologies. J. Telecommun. Digit. Econ. **9**(3), 58–86 (2021). https://doi.org/10.18080/JTDE.V9N3.418
8. D. Brenner, Global 5G spectrum update. Spectr. Strateg. Technol. Policy, 2020. https://www.qualcomm.com/content/dam/qcomm-martech/dm-assets/documents/20200625_mipi_-_5g_spectrum_-_dean_brenner_4.pdf
9. H. Tataria, M. Shafi, A.F. Molisch, M. Dohler, H. Sjoland, F. Tufvesson, 6G wireless systems: Vision, requirements, challenges, insights, and opportunities. Proc. IEEE **109**(7), 1166–1199 (2021). https://doi.org/10.1109/JPROC.2021.3061701
10. Ericsson, Cell selection and cell reselection criteria, 1999. https://www.3gpp.org/
11. M.T. Kawser, *LTE Air Interface Protocols*. (Artech House, 2011), ISBN:978-1-60807-201-9
12. Nokia: 3GPP TSG-RAN Meeting #24', Measurement Rules. vol. 7.3.2, 2004, [Online]. Available: RP-040243.pdf – 3GPP
13. 3GPP TSG RAN WG4 #2: Requirements on Handover, Agenda Item 7.4, 1999. Available: Tdoc R4-99-068. https://www.3gpp.org/

14. M. Phemina Selvi, S. Sendhilnathan, Minimizing handover delay and maximizing throughput by Heterogeneous Handover Algorithm (HHA) in telecommunication networks. Appl. Math. Inf. Sci. **11**(6), 1737–1746 (2017). https://doi.org/10.18576/amis/110621
15. K. Aswathy, P. Asok, T. Nandini, L.S. Nair, Handover latency improvement and packet loss reduction in wireless networks using scanning algorithm. Adv. Intell. Syst. Comput. **708**(March), 43–51 (2018). https://doi.org/10.1007/978-981-10-8636-6_5
16. S.K. Lee, K. Sriram, K. Kim, Y.H. Kim, N. Golmie, Vertical handoff decision algorithms for providing optimized performance in heterogeneous wireless networks. IEEE Trans. Veh. Technol. **58**(2), 865–881 (2009). https://doi.org/10.1109/TVT.2008.925301
17. S. Hara, R. Prasad, Overview of multicarrier MA, in *The Fourth IEEE International Symposium on Spread Spectrum Techniques and Applications (ISSSTA'96),* Mainz, Germany, 22–25 September 2015, pp. 126–133
18. M. Hassan, Cellular networks. Wirel. Mob. Netw., 141–163 (2022). https://doi.org/10.1201/9781003042600-11
19. X. Sun, H. Nan, N. Zheng, Study on system latency reduction based on shorten TTI. Int. Conf. Signal Process. Proc. ICSP, 1293–1297 (2016). https://doi.org/10.1109/ICSP.2016.7878035
20. H. Rohling, R. Gruenheid, Performance comparison of different multiple access schemes for the downlink of an OFDM communication system. IEEE Veh. Technol. Conf. **3**, 1365–1369 (1997). https://doi.org/10.1109/vetec.1997.605406
21. H. Xing, S. Hakola, The investigation of power control schemes for a device-to-device communication integrated into OFDMA cellular system. IEEE Int. Symp. Pers. Indoor Mob. Radio Commun. PIMRC, 1775–1780 (2010). https://doi.org/10.1109/PIMRC.2010.5671643
22. New York University: NYUSIM Millimeter-wave Channel Simulator 3.0, 2021. [Online]. Available: http://wireless.engineering.nyu.edu/nyusim
23. T.S. Rappaport, G.R. MacCartney, M.K. Samimi, S. Sun, Wideband millimeter-wave propagation measurements and channel models for future wireless communication system design. IEEE Trans. Commun. **63**(9), 3029–3056 (2015). https://doi.org/10.1109/TCOMM.2015.2434384
24. S. Sun et al., Investigation of prediction accuracy, sensitivity, and parameter stability of large-scale propagation path loss models for 5G wireless communications. IEEE Trans. Veh. Technol. **65**(5), 2843–2860 (2016). https://doi.org/10.1109/TVT.2016.2543139

# Chapter 13
# Artificial Intelligence-Based Method for Smart Manufacturing in Industrial Internet of Things Network

**Ajay Kumar Kaushik, Deepak Kumar Sharma, and Sanjay K. Dhurandher**

## 13.1 Introduction

Industrial Internet of Things (IIoT) network is designed to serve industrial needs where things are connected and giving data for analytics. Artificial Intelligence (AI) advocates a system to solve the given tasks from the learning of data/events/actions. To achieve the same, AI uses algorithms that are self-learning and intelligent. Literature shows that it might be a good idea by the merger of AI and IIoT to create artificial Industrial Internet of Things (AI-IIoT) network. Some important AI-IIoT use cases are smart manufacturing, autonomous vehicles, and smart cities.

This paper critically reviews the need, methodology, and implementation requirements in the transformation process from traditional manufacturing to smart manufacturing. AI techniques are used to improve the key manufacturing variables. For IIoT environment, Thingsworx simulator [17] is used. AI-IIoT for smart manufacturing has very stringent requirements. It is not the case of putting some logic in manufacturing processes, rather it is a systematic method to connect, collect, build, and analyze in a framework. AI techniques are then used to assist in decision-making. Result shows that AI techniques can assist in smart manufacturing in several phases and have the capabilities to improve several key process variables of product development lifecycle. The primary contributions of this paper are as follows:

A. K. Kaushik (✉) · S. K. Dhurandher
Department of Information Technology, Netaji Subhas University of Technology, Dwarka, New Delhi, India

D. K. Sharma
Department of Information Technology, Indira Gandhi Delhi Technical University for Women, Kashmere Gate, New Delhi, India

 
I. Woungang, S. K. Dhurandher (eds.), *5th International Conference on Wireless, Intelligent and Distributed Environment for Communication*, Lecture Notes on Data Engineering and Communications Technologies 174,
https://doi.org/10.1007/978-3-031-33242-5_13

1. An AI-IIoT framework for data acquisition, data transformation, and processing for rule-based decisions to improve the operations in IIoT system
2. AI-based feature selection and information gain to reduce data communication and latency

The rest of this paper is organized as follows: Sect. 13.2 contains the brief description and evaluation of the existing AI-based solutions for smart manufacturing. Section 13.3 explains the proposed model for smart manufacturing. Section 13.4 explains the results and findings for AI-based IIoT framework as a tool for smart manufacturing. Section 13.5 discusses the conclusion and future scope for the proposed work.

## 13.2 Literature Review of AI-IIoT in Smart Manufacturing

Any improvement to make the smart industry means bringing smartness in logistics, services, and supply chain [1–3]. IIoT is committed to adapt IoT to interconnect anything, anywhere, and at any time in the manufacturing systems. IIoT can improve the parameters like productivity, efficiency, safety, and intelligence with the basic IIoT architecture particularly directed toward process automation in factories. IIoT demands an increased connectivity of different components, but this kind of connectivity brings the consideration for security issues, and therefore, IIoT has stringent security requirements [3]. The contextual information collected can be used for value-added services like personalized alerts and suggestions. Wallis et al. [7] proposed a novel rule-based approach, which uses subcontexts of IIoT context for deep packet inspection to protect the privacy of data. They also developed formal definitions for IIoT context, subcontext, and contextual rules. Emil Blixt Hansen et al. [11] advocates the need for the usage of Artificial Intelligence and IoT in small- and medium-sized enterprises (SME). Large-scale enterprises have the luxury, resources, and environment to handle the complexities and challenges of AI-IIoT framework [4]. Interconnection between AI-IIoT and SME is explained in their work. Panagiotis Trakadas et al. [12] identify that data is being collected from the shop floor assets and IT systems. Data-driven analytic efforts support more informed business intelligence decisions [10, 16]. However, the analytics results obtained are being used in isolation in different phases of the production process. In their work, they proposed the full integration of AI in all parts of the manufacturing process. The data and information model in AI-IIoT Framework is bringing intelligence in smart manufacturing process. Table 13.1 summarizes the work related to AI-IIoT paradigm.

**Table 13.1** Description of some artificial intelligence(AI)-based methods for Industrial Internet of Things (IIoT)

| S. No. | Authors | Key work/idea and methodology | Conclusions by the author |
|---|---|---|---|
| 1 | Wallis et al. [7] | Security of the manufacturing data from harmful manufacturing commands. | Formal definitions of subcontext and IIoT context have been developed. |
| 2 | Loske et al. [5] | Evaluate different context awareness and authentication mechanisms. They worked on how the user authentication can be shifted toward machine authentication. | They presented overview of physical features and validation methods for preselection of several combinations and approaches. |
| 3 | Fotios Zantalis et al. [13] | Provide a review of machine learning and IoT in smart transportation. | They conclude that there is possible lack of machine learning coverage for smart lightning and smart parking. |
| 4 | Parthkumar Patel et al. [15] | Proposed a test bed for experimenting IoT applications. | They implement use cases traffic and lightning for validation purpose. |
| 5 | Devarshi Shah et al. [6] | Discussed the characteristics and requirements of big IoT data. They study the opportunities to utilize this big IoT data. | They use in house-enabled IoT ccctestbed for manufacturing. They use feature engineering as the key to develop machine learning models to determine key process identifiers. |
| 6 | Ahmad et al. [10] | Focused on IoT data analytics. They find the limitations of the existing data analytics techniques and its limitations in IoT data analytics. | Explored the requirements to integrate with AI and ML with analytics in the digital organization. |
| 7 | Emil Blixt Hansen et al. [11] | Need of AI in small- and large-scale enterprises. Discussed several enablers of AI and IoT. | Focus should be on simplified AI solutions for small- and medium-sized enterprises. |
| 8 | Panagiotis Trakadas et al. [12] | Improvised the existing layers in Industry 4.0 by human in loop. Discussed security issues. | Data and information model conversion to AI-enabled intelligence that allows insights into data for informed decisions. |
| 9 | V. Kharchenko et al. [8] | Comparison of traditional computer based manufacturing methods with digital twins. | Digital twin with AI methods in IoT environment allows breakdown prediction, optimization of maintenance schedules, etc. |

## 13.3 Proposed Approach

### *13.3.1 Motivation*

We are living in an era where the total assets of big enterprises are more than Gross Domestic Product of several countries. Customers are demanding better products and improved services. They also demand the personalized services. Enterprises are also faced with evergreen and persistent challenge that is to upgrade with respect to time and environment; hence, companies are forced to bring intelligence in their end products and services. To build enhanced relationships with customers, companies have started to upgrade toward AI-IIoT paradigm in their manufacturing plants. IIoT is supposed to increase the operational efficiencies for cost reduction. With the use of AI-based techniques, companies are able to reduce the waste elimination and to optimize the production cycle through intelligent decision-making. This is going to increase the revenues. Companies' spending on IIoT is likely to increase from $128 billion in 2020 to $411.9 billion by 2025 [16] (Fig. 13.1).

### *13.3.2 Proposed Methodology for AI-IIoT Framework*

Irrespective of the IIoT use case, an efficient AI-IIoT merger can only be successful through an architecture that facilitates to connect and understand data generated by IIoT nodes and to leverage the data with AI methods. First of all, interconnection of IIoT nodes/devices, process, and people is done. Multiple sensor nodes are attached to different IIoT nodes. Data sensed is transferred directly to cloud, or it can be given to edge nodes for intermediate processing depending upon the storage and

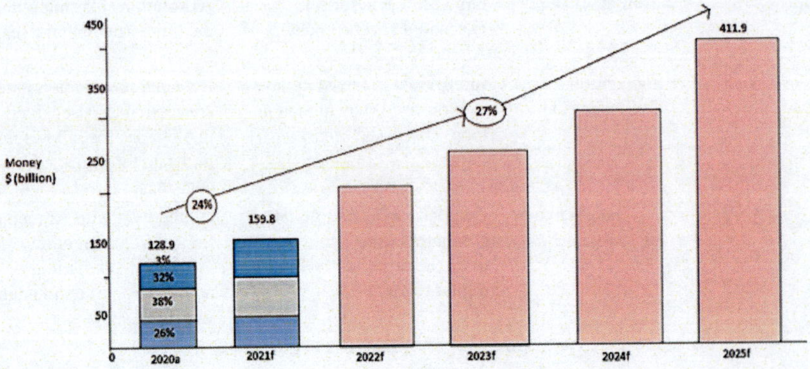


Fig. 13.1 Spending on Internet of Things (IoT) enterprise year-wise prediction [16]

computing capacity of IoT nodes. Considering the fact that industrial system will generate the vast data in continuous mode, AI-based methods are then used to deal with the complex data to extract some meaningful information out of it.

Aim of the proposed work is to prevent the engines from catastrophic failures. The goal condition is to predict the detection of low greasing condition in the motor. Model aims to use the minimum number of IIoT devices for achieving the same. Proposed model is based on the prediction of faults in the working of engines, but we can also apply this model to any manufacturing industry demanding the preventive maintenance in their operations. Further, the model can be extended to create personalized event subscription services for the customers.

### 13.3.3 Data Flow Cycle for the Proposed Model

Figure 13.2 shows the data cycle used in the proposed AI-IIoT framework for smart manufacturing. IIoT devices like sensors, actuators, conveyor belts, and robotic arms are digitally modeled using the concept of digital twins. These devices can be connected through wireless protocols like Zigbee and Bluetooth. In the proposed model, Edge Microserver Connectivity (EMS Connectivity) is used to connect the devices. EMS uses ThingWorx Always on protocol based on the Open Websocket Standard RFC6455 [17]. They are allowed to work with respect to the requirements of any use case. Data from heterogeneous devices are processed, and it is as per requirement of the use case. Now the goal condition is determined, and the AI-based feature selection is used that is correlated with the goal condition. The minimal threshold value is set to minimize the computation load by reducing the latency in decision-making. In the next step, parameters that are correlated with the goal condition are determined and fed to rule-based decision module. Decisions are sent back to IIoT environment for the continuous process improvement. Our proposed model is able to collect data, and by AI-enabled processing, it is able to select minimum sensors that can remain in active state, and others are forced into sleep state. This data flow cycle in AI-IIoT framework ensures that real-time decisions are taken continuously from the data collected for the smart manufacturing. With the ongoing preventive maintenance, we are able to deal with the unwanted situation of the catastrophic failure in the manufacturing process.

### 13.3.4 Algorithm for the Proposed Model

*Step 1:* Connect the devices like sensors and actuators through Edge Microserver Connectivity protocol to make a connected framework.

*Step 2:* Data is collected from the devices (motor), and various data quality improvement methods like data validation, data conversion, data alignment, and noise reduction are applied. In AI-IIoT based smart manufacturing, data models are

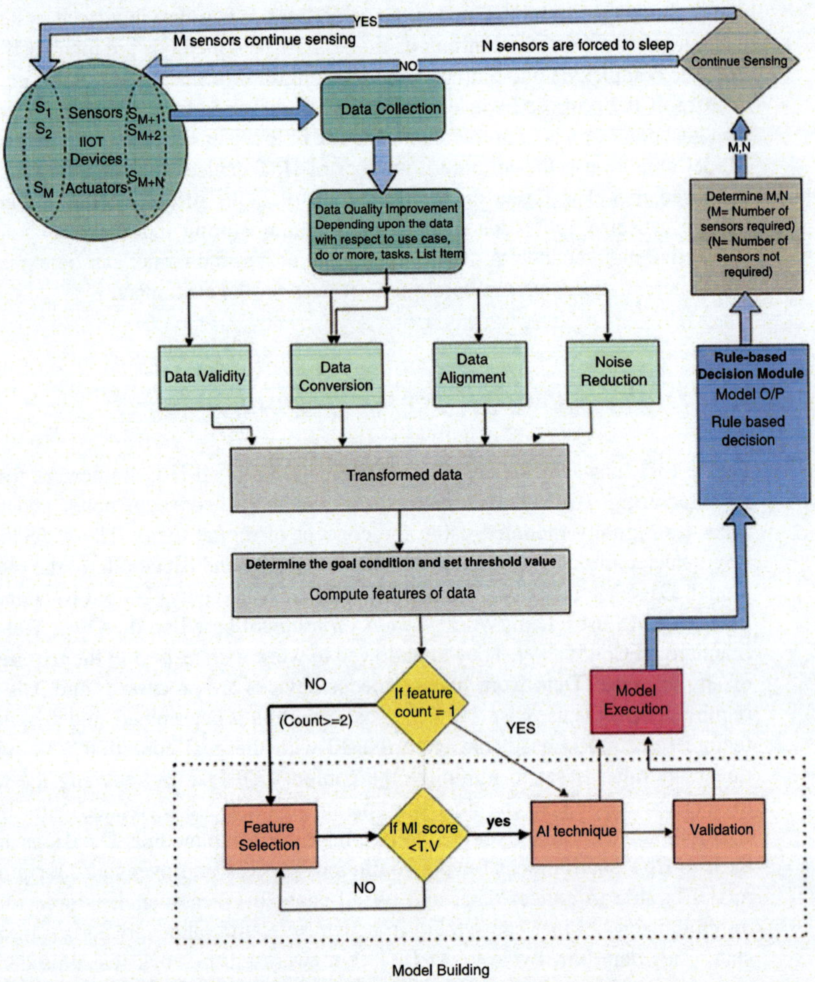


**Fig. 13.2** Data flow cycle for the proposed model

key, and a successful model demands the integration of heterogeneous data collected from different devices. Raw data is collected from multiple connected devices from IIoT platform.

*Step 3:* Output of the Step 2 is raw data, which is not useful for data classification, so the raw data is transformed through techniques like data validation, conversion, alignment, and noise reduction depending upon the use case requirements. Data labeling is done to ensure the dataset has meaningful labels so that models can learn from the dataset during training.

*Step 4:* The goal condition is determined as to detect low greasing condition in the engine motor. Goal state can be a single unique state, or it can be multiple goal states to satisfy the multiple goal conditions. Data is analyzed to identify the most relevant data through AI-based feature selection techniques. If the selected feature has mutual information score less than the threshold value, select another feature. Data is analyzed to identify the most relevant data through AI-based feature selection techniques. All the features in the dataset are computed for every signal (Sensor and Frequency band pair.) If vibration frequency is less than threshold value which is set as 0.5, goal condition is selected as 0; else it is selecteed as 1.

Dataset consists of the collected data with several features. All the features are calculated using Entropy using Eqs. (13.1) and (13.2).

$$H(X) = -\sum (pi * \log_2 pi) \tag{13.1}$$

$$\text{Entropy}(p) = -\sum\nolimits_{n=1}^{\text{N}} (pi * \log_2 pi) \tag{13.2}$$

where $H(X)$ is the heterogeneity of the events and $p$ is the entropy of the selected features.

With respect to low greasing goal condition, features are selected, which are best correlated with the target and information gain as described in Step 5.

*Step 5 Information gain:* Information gain is calculated as follows:

On the basis of Step 4 results, we create profiles by taking high correlated signals. For every profile, we compute information gain through Eq. (13.1)

$$\text{MI (Feature | Goal)} = \text{Entropy (Feature)} - \text{Entropy (Feature | Target))} \tag{13.3}$$

We compare the information gain for the selected profiles and compare with threshold value of information gain of 0.4. Only the profiles passing the threshold are taken, and the rest are discarded.

*Step 6 Model execution and rule-based decision:* Model building and validation is not enough to solve the real-world problems. In fact characteristic of the model for AI-IIoT framework is that IIoT platform provides set of data functions (collection, preprocessing, and transformation) to provide consistent data. On the other hand, the output of an AI model must combine with business logic or context to transform the insights from the model into actionable decisions. In turn, these decisions need to be acted upon, directly or indirectly, to optimize the operations of industrial processes. Model is executed to determine the strategic placement of sensors. Model calculates the prediction accuracy for various combinations of signals and profiles. This rule is sent back to IIoT environment where necessary sensors are allowed to remain in active state, and others are made to sleep state.

## 13.4 Experimental Setup and Results

### *13.4.1 Tools and Simulators Used*

Thingworx simulator [17] is used to connect, analyze, and build the different devices in this work; web edge microserver is connected with foundation. Considering the industrial requirements in mind, always on protocol is used so that there is no need to request open/forward port, and communication is established through more secured environment with server. EMS was installed and configured. EMS was connected with the foundation.

### *13.4.2 Build an Engine Analytical Model*

Motor through continuous rotation loses grease and may fail catastrophically if they are not properly serviced. To prevent such failures, motors are fitted with sensors to record vibration. Expectation from the proposed model is that motors can detect certain vibrations indicating that preventive maintenance is due, so the key is to monitor and record vibration data from the prototype motor. Mutual information is computed for every Si_Fbj combination. Mutual information is zero for those Si_Fbj values, which have same base value between a good grease condition and a bad grease condition. Models are primarily used by analytics manager, but they can still be used to define accuracy of predictions. Prediction model is then run against withheld data. This forms an accuracy measure. Withhold data is selected randomly to get a different model and accuracy measure each time when a model is created on the same dataset.

### *13.4.3 Evaluation of the Proposed Model*

The main aim is to prevent the engine from failing. For this, vibration sensing was monitored. Some of the applications of vibration sensing are to monitor the rotating equipment like gas turbines, compressors, and rolling mills. Many industries like oil and gas, refining, and petrochemicals employ vibration sensors. A vibration sensor measures the frequency and quantity of vibrations. This metric helps the maintenance teams to monitor the conditions that lead to equipment failure. Working principle of a vibration sensor is to measure the velocity changes in the device. Vibration data was collected to create an analytical model.

In the proposed methodology, devices in the production line should send data continuously for analytics. Such kind of data flow from devices to analytical managers results in creating event subscription services. In this work, we are able to create alerts for preventive maintenance through greasing alerts in abnormal conditions. In smart manufacturing Edge Microserver (EMS) readings, script can be tagged on the device to analyze the variables in real time. This allows event subscription model for customers to notify various alerts as per the properties set up in the even subscription model. Algorithm computes the mutual information and discards the frequency bands from the vibration sensors that are not able to detect noise.

### 13.4.4 Result Analysis and Discussion

Sensors were placed to collect the data at different frequency bands. In first iteration, we take only two sensors with five frequency bands and the threshold value of Mutual Information (MI) as 0.5.

Table 13.2 describes the following fields:

1. Si_Fbj combination, which indicates the vibrations are measured from the Frequency band Fbj for Sensor Si.
2. Mutual information score, which is an attempt to measure the heterogeneity or uncertainty within the sensor readings. There is a relationship between the heterogeneity and the uncertainty; the more heterogeneous event means more uncertainty. Shannon expressed the relationship between probability, heterogeneity, and impurity.
3. Signal name is the parameter selected. Model works on the vibration frequencies measured in the manufacturing unit.

Table 13.2 shows that all the frequency bands of sensor 1 are able to detect low greasing conditions with MI > 0.5. On the other hand, only the first and second frequency bands of sensor 2 are able to detect low greasing conditions with MI > 0.5. Next, we make the profile by picking the most correlated frequency bands from Table 13.5.

**Table 13.2** Sensor readings

| S. No. | Feature name | Mutual information | Signal_name | Dataset | Goal |
|---|---|---|---|---|---|
| 1. | S1_fb1 | 0.06 | Vibration_frequency | Vibration_Dataset | Low_Grease_Condition |
| 2. | S1_fb2 | 0.07 | | | |
| 3. | S1_fb3 | 0.04 | | | |
| 4. | S1_fb4 | 0.06 | | | |
| 5. | S1_fb5 | 0.07 | | | |
| 6. | S2_fb1 | 0.05 | | | |
| 7. | S2_fb2 | 0.05 | | | |
| 8. | S2_fb3 | 0.03 | | | |
| 9. | S2_fb4 | 0.04 | | | |
| 10. | S2_fb5 | 0.04 | | | |

Next task was to explore with the combination of variables into profiles. We create profiles by combining data from those frequency bands that are highly correlated with the goal.

For this, comparison was done through *Z* scores. *Z* score, which is also known as standard score, gives an idea how far from the mean a data point is.

$$Z = \frac{X - \mu}{\sigma} \tag{13.4}$$

where $X$ is Score, $\mu$ is Mean, and $\sigma$ is Standard Deviation.

Table 13.3 evaluates different profiles. Profiles are combination of several Si_Fbj combinations. It was calculated for every Si_Fbj combination. However, it was observed that only three profiles consists of significant number of records. Values from Table 13.6 indicate that profile 2 and profile 3 are more correlated with the goal condition. Profile with MI $<$ threshold value 0.5 were discarded.

Model was simulated using an analytics manager. Information gain is calculated by comparing the entropy of the dataset before and after a transformation. Information gain is used for feature selection, by computing gain of all variables with respect to target signal. Profiles are calculated through *Z* score.

Tables 13.4 and 13.5 evaluate the model accuracy using Confusion matrix. Model results are compared with the withheld data. It was observed that when sensor 1 was used alone, accuracy for true positive and true negatives were 25.125% and 49%, respectively. When it was used in combination with sensor 2, accuracy was 26.625% and 49.875%. Frequency bands for sensor 1 are highly correlated than bands for sensor 2. Sensor 1 is more significant than sensor 2. It leaves to the decision-makers whether to leave sensor 2 or to be used in combination with sensor 1.

### 13.4.5 Results on Varying Number of Sensors

To determine the optimum number of sensors required in the manufacturing unit, we collect the data from five sensors.

Tables 13.8 and 13.9 evaluate the model accuracy using Confusion matrix. It was observed that when sensor 1 was used alone, accuracy for true positive and true negatives were 26.5% and 49.7%, respectively. When it was used in combination with sensors 2–5, accuracy was 27.8% and 50.1%, respectively. Frequency bands for sensor 1 are highly correlated than bands for sensors 2–5. Sensor 1 is more significant than combination of sensor 2–5. It leaves to the decision-makers whether to leave sensor 2–5 or to be used in combination with sensor 1. This shows that sensors 2–5 can be forced to sleep while sensor 1 is allowed to continue sensing.

**Table 13.3** Evaluating profiles

| S. No. | Feature name | Vibration values | Percentage of records | Signal_name | Average goal | Goal | Filter | Average value for goal |
|---|---|---|---|---|---|---|---|---|
| *Profile no. 1* | | | | | | | | |
| 1. | S1_fb1 | 81–110 | 18.39 | Vibration_frequency | .48 | Low_Grease_Condition | Threshold value | 0.40 |
| And | S1_fb4 | 101–178 | 11.67 | | .37 | | | |
| And | S2_fb1 | 155–189 | 16.55 | | .43 | | | |
| And | S2_fb2 | 98–178 | 13.34 | | .45 | | | |
| *Profile no. 2* | | | | | | | | |
| 1. | S1_fb2 | 101–211 | 10.02 | Vibration_frequency | .48 | Low_Grease_Condition | Threshold value | 0.40 |
| And | S1_fb1 | 153–235 | 12.08 | | .47 | | | |
| And | S2_fb1 | 156–198 | 9.07 | | .42 | | | |
| And | S2_fb2 | 211–228 | 14.56 | | .46 | | | |
| *Profile no. 3* | | | | | | | | |
| 1. | S1_fb1 | 168–218 | 15.03 | Vibration_frequency | .47 | Low_Grease_Condition | Threshold value | 0.40 |
| And | S1_fb5 | 193–228 | 14.87 | | .46 | | | |
| And | S2_fb2 | 221–256 | 14.67 | | .43 | | | |
| And | S2_fb1 | 164–208 | 13.80 | | .45 | | | |

**Table 13.4** Confusion matrix for sensor 1 alone

| | |
|---|---|
| *Accurate*<br>No. of records: 201<br>% of reading: 25.125% | *False positive*<br>No. of records: 65<br>% of reading: 8.125% |
| *False negative*<br>No. of records: 142<br>% of reading: 17.75% | *Accurate*<br>No. of records: 392<br>% of reading: 49% |
| True | False |

**Table 13.5** Confusion matrix for sensor 1 and sensor 2

| | |
|---|---|
| *Accurate*<br>No. of records: 213<br>% of reading: 26.625% | *False positive*<br>No. of records: 61<br>% of reading: 7.625% |
| *False negative*<br>No. of records: 135<br>% of reading: 16.875% | *Accurate*<br>No. of records: 399<br>% of reading: 49.875% |
| True | False |

### *13.4.6 Results on Varying Number of Records*

In the next iteration, we take more number of sensor readings to find the result when numbers of records are increased.

Tables 13.10 and 13.11 evaluate the model accuracy as done previously with more number of records. It was observed that when sensor 1 was used alone with 30,000 records, accuracy for true positive and true negatives were 23.49% and 65.15%, respectively. When it was used with 800 records, accuracy was 26.5% and 49.7%. The results prove that when more data is supplied, accuracy increased to 89% from 76%.

It was observed that when sensor 1 in combination with sensor 2–5 was used alone with 30,000 records, accuracy for true positive and true negatives were 24.18% and 67.5%, respectively. When it was used with 800 records, accuracy for true positives and true negatives were 27.8% and 50.1%, respectively. The results prove that when more correlated data is supplied, accuracy increased to 91.68% from 77.9%.

Our analysis shows that results from Tables 13.2, 13.3, 13.4, 13.5, 13.6, 13.7, 13.8, 13.9, 13.10, and 13.11 suggest that more number of sensors is not typically required. Figure 13.3 shows the same efficiency (approximate) can be achieved if more data from the sensor with most correlated frequency bands is taken. AI-based feature selection ensures that efficiency increases with more number of records. It allows some of the sensors, which are not strongly correlated with the goal condition to remain in sleep mode, whereas sensors with strongly correlated sensors are allowed to be in active state.

**Table 13.6** Sensor readings for five sensors

| S. No. | Feature name | Mutual information | Signal_Name | Dataset | Goal |
|---|---|---|---|---|---|
| 1. | S1_fb1 | 0.07 | Vibration_frequency | Vibration_Dataset | Low_Grease_Condition |
| 2. | S1_fb2 | 0.06 | | | |
| 3. | S1_fb3 | 0.08 | | | |
| 4. | S1_fb4 | 0.06 | | | |
| 5. | S1_fb5 | 0.08 | | | |
| 6. | S2_fb1 | 0.04 | | | |
| 7. | S2_fb2 | 0.05 | | | |
| 8. | S2_fb3 | 0.04 | | | |
| 9. | S2_fb4 | 0.03 | | | |
| 10. | S2_fb5 | 0.05 | | | |
| 11. | S3_fb1 | 0.05 | | | |
| 12. | S3_fb2 | 0.04 | | | |
| 13. | S3_fb3 | 0.06 | | | |
| 14. | S3_fb4 | 0.05 | | | |
| 15. | S3_fb5 | 0.06 | | | |
| 16. | S4_fb1 | 0.05 | | | |
| 17. | S4_fb2 | 0.04 | | | |
| 18. | S4_fb3 | 0.03 | | | |
| 19. | S4_fb4 | 0.04 | | | |
| 20. | S4_fb5 | 0.05 | | | |
| 21 . | S5_fb1 | 0.06 | | | |
| 22. | S5_fb2 | 0.06 | | | |
| 23. | S5_fb3 | 0.04 | | | |
| 24. | S5_fb4 | 0.05 | | | |
| 25. | S5_fb5 | 0.06 | | | |

## 13.5 Conclusion and Future Scope

In this work, the proposed model integrates the IIoT devices with the AI-enabled data processing model. AI model is executed and is validated against the withheld data. Proposed work successfully makes a framework for AI-IIoT-enabled manufacturing. Proposed model helps in the prediction of conditions leading to the catastrophic failure of the engines due to the low greasing condition. Proposed model shows that when more number of sensors are used in comparison to single sensor efficiency is not improved much. However, when single sensor is supplied with more records, proposed model selects the records with better efficiency due to the AI-based techniques like feature selection and information gain. Rule-based decisions are sent back to IIoT environment, which increases the prediction efficiency for the next iterations. The proposed model can be extended to provide the personalized event/ subscription services in manufacturing industries. AI-IIoT-

**Table 13.7** Evaluating profiles

| S. No. | Feature name | Vibration values | Percentage of records | Signal_name | Average goal | Goal | Filter | Average value for goal |
|---|---|---|---|---|---|---|---|---|
| *Profile no. 1* | | | | | | | | |
| 1. | S1_fb1 | 89–120 | 19.69 | Vibration_frequency | .43 | Low_Grease_Condition | Threshold value | 0.40 |
| And | S2_fb2 | 110–189 | 14.56 | | .46 | | | |
| And | S3_fb4 | 159–196 | 18.69 | | .41 | | | |
| And | S5_fb2 | 101–187 | 16.56 | | .44 | | | |
| *Profile no. 2* | | | | | | | | |
| 1. | S1_fb2 | 110–210 | 14.27 | Vibration_frequency | .46 | Low_Grease_Condition | Threshold value | 0.40 |
| And | S1_fb5 | 169–235 | 15.18 | | .39 | | | |
| And | S5_fb4 | 189–220 | 13.10 | | .47 | | | |
| And | S3_fb5 | 215–234 | 16.67 | | .45 | | | |
| *Profile no. 3* | | | | | | | | |
| 1. | S1_fb3 | 198–228 | 16.39 | Vibration_frequency | .46 | Low_Grease_Condition | Threshold value | 0.40 |
| And | S1_fb4 | 197–238 | 18.78 | | .49 | | | |
| And | S3_fb1 | 215–236 | 16.76 | | .45 | | | |
| And | S5_fb2 | 179–223 | 17.89 | | .42 | | | |

**Table 13.8** Confusion matrix for Sensor 1 alone

| | |
|---|---|
| *Accurate*<br>No. of records: 212<br>% of reading: 26.5% | *False positive*<br>No. of records: 45<br>% of reading: 5.6% |
| *False negative*<br>No. of records: 145<br>% of reading: 18.1% | *Accurate*<br>No. of records: 398<br>% of reading: 49.7% |
| True | False |

**Table 13.9** Confusion matrix for sensors 1–5

| | |
|---|---|
| *Accurate*<br>No. of records: 223<br>% of reading: 27.8% | *False positive*<br>No. of records: 54<br>% of reading: 6.7% |
| *False negative*<br>No. of records: 122<br>% of reading: 15.2% | *Accurate*<br>No. of records: 401<br>% of reading: 50.1% |
| True | False |

**Table 13.10** Confusion matrix for sensor 1 alone

| | |
|---|---|
| *Accurate*<br>No. of records: 7048<br>% of reading: 23.49% | *False positive*<br>No. of records: 899<br>% of reading: 2.9% |
| *False negative*<br>No. of records: 2597<br>% of reading: 8.65% | *Accurate*<br>No. of records: 19,546<br>% of reading: 65.15% |
| True | False |

**Table 13.11** Confusion matrix for sensors 1–5

| | |
|---|---|
| *Accurate*<br>No. of records: 7256<br>% of reading: 24.18% | *False positive*<br>No. of records: 819<br>% of reading: 2.7% |
| *False negative*<br>No. of records: 1671<br>% of reading: 5.5% | *Accurate*<br>No. of records: 20,254<br>% of reading: 67.5% |
| True | False |

enabled smart manufacturing allows the monitoring of key process variables in hazardous conditions where it is difficult to detect the conditions leading toward the failure of the key process variables and operations.

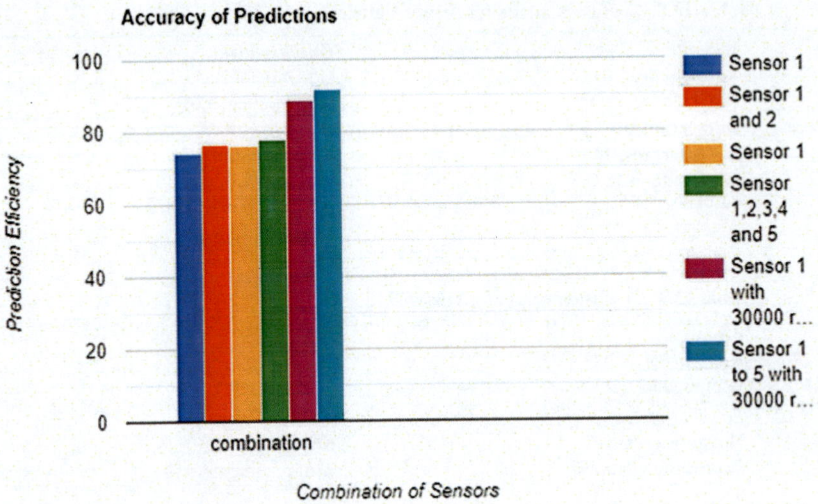


**Fig. 13.3** Prediction efficiency

# References


1. A. Colakovic, M. Hadzialic, Internet of Things (IoT): A review of enabling technologies, challenges, and open research issues. Comput. Netw. **144**, 17–39, . ISSN 1389-1286 (2018). https://doi.org/10.1016/j.comnet.2018.07.017
2. C. Perera et al., Semantic-driven configuration of Internet of Things middleware, in *The 9th International Conference on Semantics, Knowledge and Grids.*, (IEEE, 2013), pp. 66–73
3. K. Tange, M. De Donno, X. Fafoutis, N. Dragoni, A systematic survey of industrial Internet of Things security: Requirements and fog computing opportunities. IEEE Commun. Surv. Tutorials **22**(4), 2489–2520 (2020). https://doi.org/10.1109/COMST.2020.3011208
4. J. Gubbi et al., Internet of Things (IoT): A vision, architectural elements, and Future directions. Futur. Gener. Comput. Syst. **29**, 7 (2013)
5. M. Loske, L. Rothe, D.G. Gertler, Context-aware authentication:- State-of-the-art evaluation and adaptation to the IIoT, in *2019 IEEE 5th World Forum on Internet of Things (WF-IoT)*, (2019), pp. 64–69. https://doi.org/10.1109/WF-IoT.2019.8767327
6. D. Shah, J. Wang, Q. He, Feature engineering in big data analytics for IoT-enabled smart manufacturing – Comparison between deep learning and statistical learning. Comput. Chem. Eng. **141**, 106970 (2020). https://doi.org/10.1016/j.compchemeng.2020.106970
7. K. Wallis, M. Hüffmeyer, A.S. Koca, C. Reich, *Access Rules Enhanced by Dynamic IIoT Context* (2018), pp. 204–211. https://doi.org/10.5220/0006688502040211
8. V. Kharchenko, O. Illiashenko, O. Morozova, S. Sokolov, Combination of digital twin and artificial intelligence in manufacturing using industrial IoT, in *2020 IEEE 11th International Conference on Dependable Systems, Services and Technologies (DESSERT)*, 2020, pp. 196–201. https://doi.org/10.1109/DESSERT50317.2020.9125038; A. Botta et al., Integration of cloud computing and Internet of Things: A survey. Futur. Gener. Comput. Syst. **56**(3), 684–700 (2016)
9. K. Alexopoulos, K. Sipsas, E. Xanthakis, S. Makris, D. Mourtzis, An industrial Internet of things based platform for context-aware information services in manufacturing. Int. J. Comput. Integr. Manuf., 1–13 (2018). https://doi.org/10.1080/0951192X.2018.1500716

10. S. Ahmad, S. Miskon, *IoT Driven Resiliency with Artificial Intelligence, Machine Learning and Analytics for Digital Transformation* (2020), pp. 1205–1208. https://doi.org/10.1109/DASA51403.2020.9317177
11. E. Hansen, S. Bogh, Artificial intelligence and internet of things in small and medium-sized enterprises: A survey. J. Manuf. Syst. **58** (2020). https://doi.org/10.1016/j.jmsy.2020.08.009
12. P. Trakadas, P. Simoens, P. Gkonis, An artificial intelligence-based collaboration approach in industrial IoT manufacturing: Key concepts, architectural extensions and potential applications. Electronics **20**, 5480 (2020). https://doi.org/10.3390/s20195480
13. F. Zantalis, G. Koulours, S. Karebetos, D. Kandris, A review of machine learning and IoT in smart transportation. Open Access, MDPI. https://doi.org/10.3390/fi11040094
14. G. Manogaran, R. Varatharajan, D. Lopez, P.M. Kumar, R. Sundarasekar, C. Thota, A new architecture of Internet of Things and big data ecosystem for secured smart healthcare monitoring and alerting system. Future Gener. Comput. Syst. **82**, 375–387 (2018)
15. P. Patel, J. Dave, S. Dalal, P. Patel, S. Chaudhary, A testbed for implementing IoT applications, https://arxiv.org/abs/1705.07848
16. P. Wegner, In the global IoT spending (2021, July 2016). Retrieved from http://iot-analytics.com/2021-global-iot-spending-grow
17. https://www.ptc.com/en/products/developer-tool

# Index




I. Woungang, S. K. Dhurandher (eds.), *5th International Conference on Wireless, Intelligent and Distributed Environment for Communication*, Lecture Notes on Data Engineering and Communications Technologies 174,
https://doi.org/10.1007/978-3-031-33242-5